# SpringerBriefs in Electrical and Computer Engineering

For further volumes:
http://www.springer.com/series/10059

Vikram Arkalgud Chandrasetty

# VLSI Design

## A Practical Guide for FPGA and ASIC Implementations

 Springer

Vikram Arkalgud Chandrasetty
University of South Australia
Adelaide, Australia
vikramac@ieee.org

ISSN 2191-8112             e-ISSN 2191-8120
ISBN 978-1-4614-1119-2     e-ISBN 978-1-4614-1120-8
DOI 10.1007/978-1-4614-1120-8
Springer New York Dordrecht Heidelberg London

Library of Congress Control Number: 2011934747

Printed on acid-free paper

Springer is part of Springer Science+Business Media (www.springer.com)

*To*
*My Family and Friends*

# Preface

The area of VLSI design has gained enormous popularity over the past few decades due to the rapid advancements in integrated circuit (IC) design and technology. The ability to produce miniaturized circuits with high performance in terms of power and speed is the reason for its popularity. Low production cost and advanced techniques for reduced time-to-market adds to the ever-growing demand for ICs. The two major IC design flows – FPGA and ASIC have their own advantages and disadvantages. FPGAs are widely used for quick prototyping and also implementation of various multimedia applications by compromising power, area and speed performance with substantially reduced time-to-market and cost factors. Using ASIC technology, it has been possible to develop high performance multi-core processors. Verification and testing of such complex designs is a critical and challenging task to ensure the quality of the resulting circuits. The advances in EDA software and CAD tools alleviate the effort necessary to carry out the cumbersome design and verification process of ICs.

As we understand that the subject of VLSI design is vast, it is quite complex to find and comprehend the complete details about the design process. This book *VLSI Design: A practical guide for FPGA and ASIC implementations* provides an insight into practical design of VLSI circuits with minimal theoretical arguments. While this publication is not a complete text book on VLSI design, it is intended to serve as supplementary or reference material on practical design and implementation of VLSI circuits. The content of the book is focused for novice VLSI designers and other enthusiasts who would like to understand the VLSI practical design flows. The designs are demonstrated using industry standard software from MATLAB®, Mentor Graphics®, Xilinx®, Synopsys® and Cadence®.

I encourage you to send any errata or feedback for improving the quality of this book to *vikramac@ieee.org*.
Thank you,

Adelaide, Australia                                           Vikram Arkalgud Chandrasetty

# Contents

**1  CMOS Digital Design** ............................................................................. 1
   1.1  Design of CMOS SRAM Cell and Array ............................................. 1
       1.1.1  Plan of SRAM Cell and Array ..................................................... 1
       1.1.2  Design of 6 Transistor SRAM Cell .............................................. 2
       1.1.3  Simulations of SRAM Cell ......................................................... 2
       1.1.4  Layout of SRAM Cell .................................................................. 3
       1.1.5  Design of SRAM Array ............................................................... 4
       1.1.6  Simulation of SRAM Array ........................................................ 4
   1.2  Design of SRAM Chip Circuit Elements ............................................ 5
       1.2.1  SRAM Chip Circuit Elements .................................................... 5
       1.2.2  Design of Complete SRAM Chip ................................................ 8
       1.2.3  Simulations of Complete SRAM Chip ....................................... 10
       1.2.4  Delay Extraction for SRAM Chip Write/Read
            Operation .................................................................................... 10
       1.2.5  Re-Design of SRAM Chip for Low Power
            Consumption ............................................................................... 10
   Appendix ............................................................................................... 12
   References ............................................................................................. 15

**2  FPGA Application Design** ..................................................................... 17
   2.1  Design of Direct Sequence-Spread Spectrum System ...................... 18
       2.1.1  PN Sequence Generator .............................................................. 18
       2.1.2  Transmitter for Direct Sequence-Spread
            Spectrum System ........................................................................ 21
       2.1.3  Receiver for Direct Sequence-Spread
            Spectrum System ........................................................................ 24
   2.2  FIR Filter Design ............................................................................... 29
       2.2.1  Concepts of FIR Filter ............................................................... 29
       2.2.2  Low Pass FIR Filter Design ....................................................... 30
       2.2.3  Distributed Arithmetic Architecture .......................................... 31
       2.2.4  Simulation and Synthesis Results .............................................. 31

2.3    Discrete Cosine Transform Algorithms ............................................. 32
       2.3.1    Concepts of DCT ...................................................... 32
       2.3.2    DCT Architectures on FPGA.............................................. 33
       2.3.3    Scaled 1-D 8-Point DCT Architecture ................................... 34
       2.3.4    Simulation and Synthesis Results ..................................... 35
2.4    Convolution Codes and Viterbi Decoding ..................................... 36
       2.4.1    Concepts of Convolution Codes......................................... 36
       2.4.2    Viterbi Decoder....................................................... 38
       2.4.3    Simulation and Synthesis Results ..................................... 40
Appendix.............................................................................. 42
References............................................................................ 46

3   ASIC Design ....................................................................... 47
3.1    ASIC Front-End Memory Design................................................... 47
       3.1.1    Introduction.......................................................... 47
       3.1.2    Memory Architecture and Specifications................................ 48
       3.1.3    Implementation and Simulations ...................................... 48
       3.1.4    Results Analysis and Conclusion...................................... 49
3.2    ASIC Front-End Matrix Multiplier Design........................................ 51
       3.2.1    Introduction.......................................................... 51
       3.2.2    Problem Statement..................................................... 52
       3.2.3    Matrix Multiplier Design ............................................. 52
       3.2.4    Implementation and Simulations ...................................... 52
       3.2.5    Analysis of Results and Conclusion .................................. 54
3.3    Physical Design of Matrix Multiplier .......................................... 57
       3.3.1    Introduction to Systolic Array Matrix Multiplier ................... 57
       3.3.2    Physical Design Flow.................................................. 59
       3.3.3    Results and Conclusion................................................ 78
Appendix.............................................................................. 79
References............................................................................ 81

4   Analog and Mixed Signal Design................................................... 83
4.1    Schematic Design of OPAMP...................................................... 83
       4.1.1    Introduction.......................................................... 83
       4.1.2    Two Stage OPAMP Design................................................ 84
       4.1.3    Results............................................................... 93
4.2    Layout Design of OPAMP ........................................................ 93
       4.2.1    Introduction.......................................................... 93
       4.2.2    Layout Design......................................................... 93
       4.2.3    Summary and Results .................................................. 98
Appendix.............................................................................. 99
References............................................................................ 104

About the Author ..................................................................... 105

# Abbreviations

| | |
|---|---|
| ADC | Analog to Digital Converter |
| ASIC | Application Specific Integrated Circuit |
| ATM | Asynchronous Transfer Mode |
| AWGN | Additive White Gaussian Noise |
| | |
| BJT | Bipolar Junction Transistor |
| BPSK | Binary Phase Shift Keying |
| | |
| CAD | Computer Aided Design |
| CDMA | Code Division Multiple Access |
| CDR | Clock Data Recovery |
| CMOS | Complementary Metal Oxide Semiconductor |
| CORDIC | Coordinate Rotation Digital Computer |
| CP | Charge Pump |
| CTO | Clock Tree Optimization |
| CTS | Clock Tree Synthesis |
| | |
| DAA | Distributed Arithmetic Architecture |
| DAC | Digital to Analog Converter |
| DCT | Discrete Cosine Transform |
| DEF | Design Exchange Format |
| DFM | Design For Manufacturability |
| DFT | Design For Testability |
| DRAM | Dynamic Random Access Memory |
| DRC | Design Rule Check |
| DSPF | Detailed Standard Parasitic Format |
| DSSS | Direct Sequence Spread Spectrum |
| DTC | Divide by Two Circuit |
| DTFS | Deflash Trim Form Singulation |
| DUT | Device Under Test |
| DWT | Discrete Wavelet Transform |

EDA           Electronic Design Automation
EEPROM        Electrically Erasable Programmable Read Only Memory
ERC           Electrical Rule Check

FDA           Functional Data Analysis
FEC           Forward Error Correction Codes
FF            Flip Flop
FFT           Fast Fourier Transform
FIR           Finite Impulse Response
FPGA          Field Programmable Gate Array
FSM           Finite State Machine

GDS II        Graphic Data System II
GUI           Graphical User Interface

HDL           Hardware Description Language

ICMR          Input Common Mode Range
IGFET         Insulated Gate Field Effect Transistor
IOV           Input Offset Voltage
ITF           Interconnect Technology Format
ITRS          International Road Map for Semiconductors

JFET          Junction Field Effect Transistor
JPEG          Joint Photographic Experts Group

LEF           Library Exchange Format
LFSR          Linear Feedback Shift Register
LP            Low Pass
LPE           Layout Parasitic Extraction
LSB           Least Significant Bit
LUT           Look Up Table
LVS           Layout Versus Schematic

MAC           Multiply And Accumulate
MBE           Molecular Beam Epitaxy
MEMS          Mico Electro Mechanical System
MOSFET        Metal Oxide Semiconductor Field Effect Transistor
MOSIS         Metal Oxide Semiconductor Implementation Service
MPEG          Moving Picture Experts Group
MSB           Most Significant Bit

OVS           Output Voltage Swing

PFD           Phase Frequency Detector
PG            Power Ground
PIT           Progressive Image Transmission
PLL           Phase Locked Loop
PN            Pseudo-random Noise
PPO           Post Placement Optimization
PWM           Pulse Width Modulation

| | |
|---|---|
| QAM | Quadrature Amplitude Modulation |
| QDR | Quad Data Rate |
| QPSK | Quadrature Phase Shift Keying |
| RC | Resistance Capacitance |
| RF | Radio Frequency |
| ROM | Read Only Memory |
| RTL | Register Transfer Level |
| SDC | Synopsys Design Constraint |
| SDF | Standard Delay Format |
| SNR | Signal to Noise Ratio |
| SOI | Silicon On Insulator |
| SOP | Sum Of Products |
| SPEF | Standard Parasitic Exchange Format |
| SRAM | Static Random Access Memory |
| STA | Static Timing Analysis |
| TDF | Top Design Format |
| TLU | Table Look Up |
| TSMC | Taiwan Semiconductor Manufacturing Company |
| TS-OPAMP | Two Stage–Operational Amplifier |
| USB | Universal Serial Bus |
| VCD | Value Change Dump |
| VCO | Voltage Controlled Oscillator |

# Chapter 1
# CMOS Digital Design

The demand for electronic and multimedia devices is increasing exponentially. This demand in-turn has propelled the need for memory chips to process instructions, store data and other multimedia content. Some of the most common memory structures used for faster data and program memory access are Static (SRAM) and Dynamic (DRAM) memory.

In this chapter, a 6 Transistor CMOS based SRAM memory chip of 1 KB capacity is designed and simulated. The complete chip along with SRAM cells array and circuit elements is designed using SPICE program. Simulations for the design are done using LTspice. Schematic and Layout for a single SRAM cell is also designed using Cadence Schematic and Virtuoso tool respectively. An estimation of parasitic resistance and capacitance values for the layout drawn for the SRAM cell is extracted Vituoso.

The prerequisite to approach this chapter would be an adequate background of CMOS digital circuits, Spice programming and basic knowledge of IC layout design.

## 1.1 Design of CMOS SRAM Cell and Array

### 1.1.1 Plan of SRAM Cell and Array

Static Random Access Memory (SRAM) is a type of semiconductor memory. The word static indicates that the memory retains its contents as long as power remains applied. 'Random Access' means that the location in the memory can be written to or read from in any order regardless of the memory location that was last accessed [1]. The SRAM cell has the capability to store one bit data as long as the power is continuously applied. Hence SRAM's are volatile memory devices. An array of eight SRAM cells can store 1 byte of data. Considering this unit of 8-bit SRAM array, a number of these structures can be replicated to build a large memory block. In this chapter, an SRAM memory of 1 KB is designed using 6 Transistor (6T) CMOS SRAM cell.

V.A. Chandrasetty, *VLSI Design: A Practical Guide for FPGA and ASIC Implementations*, SpringerBriefs in Electrical and Computer Engineering, DOI 10.1007/978-1-4614-1120-8_1, © Springer Science+Business Media, LLC 2011

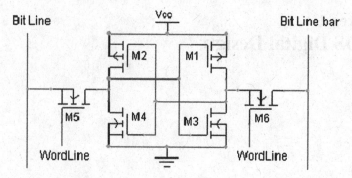

**Fig. 1.1**  6T CMOS SRAM cell

## 1.1.2   Design of 6 Transistor SRAM Cell

An SRAM cell can store one bit data on four transistors that form two cross-coupled inverters [1]. This storage cell has two stable states which are used to denote 0 and 1. Two additional access transistors serve to control the access to storage cell during read and write operations. Thus, a combination of 6 transistors is used to store one bit data.

Access to the cell is enabled by the word line (WL) which controls the two access transistors. They are used to transfer data for both read and write operations by connecting the bit lines (BL and BL bar). Although the two bit lines are not necessary, both the signal and its inverse are typically provided since it improves noise margins. The symmetric structure of SRAM's allows for differential signaling, which makes small voltage swings more easily detectable. A schematic of 6T CMOS SRAM cell is shown in Fig. 1.1.

## 1.1.3   Simulations of SRAM Cell

An SRAM cell has three different states of operation: *standby* when the circuit is idle, *reading* when the data has been requested and *writing* when updating the contents. Each states are discussed with respect to the Fig. 1.1 as follows [2]:

**Standby:**
If the word line is not asserted, the access transistors M5 and M6 disconnect the cell from bit lines. The two cross coupled inverters formed by M1–M4 will continue to reinforce each other as long as they are disconnected from the outside world.

**Reading:**
Assuming that the content of memory is 1, when the word line is asserted, the state stored in the cell is transferred to the bit line which is then read on the data output port. If the memory content was a 0, the opposite would happen.

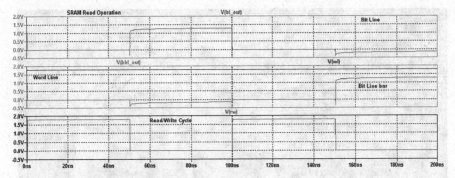

**Fig. 1.2** Spice simulations for SRAM cell read operation

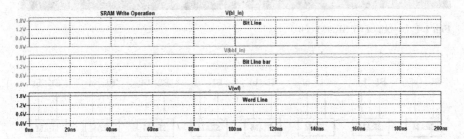

**Fig. 1.3** Spice simulations for SRAM cell write operation

*Writing*:

The start of a write cycle begins by applying the value to be written to the bit lines. The word line is asserted to store the input data on to the cell. The bit line input drivers are designed to be much stronger than the relatively weak transistors in the cell itself, so that they can easily override the previous state of the cross-coupled inverters. Careful sizing of the transistors in the SRAM cell is needed to ensure proper operation.

The spice simulation for SRAM cell Read operation is shown in Fig. 1.2.

The spice simulation for SRAM cell Write operation is shown in Fig. 1.3.

### 1.1.4 Layout of SRAM Cell

The layout for SRAM cell is drawn using Cadence Virtuoso for 180 nm technology. The layout is successfully completed with Design Rule Checks (DRC) and Layout versus Schematic (LVS) evaluation as well. A snapshot of the layout of SRAM cell is shown in Fig. 1.4. The resistance and capacitance parasitic parameters are extracted from the layout using Cadence Virtuoso.

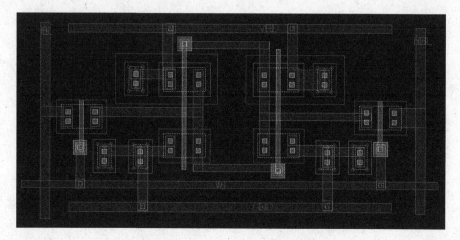

**Fig. 1.4** Layout of 6T SRAM cell

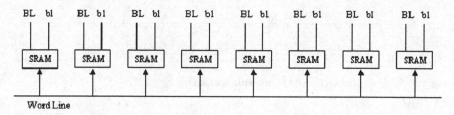

**Fig. 1.5** Block diagram of 8-bit SRAM array

## 1.1.5   Design of SRAM Array

For this design, 1 KB SRAM chip with 8-bit data I/O is required. Since one bit data can be stored in a single SRAM cell, an array of 8 cells should satisfy the requirement. Hence 1,024 such arrays are required to build 1 KB memory chip. A block diagram representing an 8-bit memory SRAM array is shown in Fig. 1.5.

## 1.1.6   Simulation of SRAM Array

An SRAM array (8-bit) is selected or activated by the row and column decoder based on the input address. The spice simulation for SRAM array section is shown in Fig. 1.6.

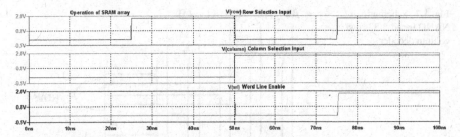

**Fig. 1.6**  Spice simulations for section of SRAM array

## 1.2   Design of SRAM Chip Circuit Elements

### 1.2.1   SRAM Chip Circuit Elements

The 6T CMOS SRAM chip requires various circuit elements to execute the desired memory operations. In this section, a complete SRAM chip circuitry elements such as, address decoder, sense amplifier, pre-charge circuit and data I/O control logic is designed using LTspice. The detailed design, schematic and simulation of these circuit elements are discussed in the following sections:

#### 1.2.1.1   Address Decoder

The Address Decoder is nothing but a simple logic circuitry used to select and enable the memory cells in the SRAM array corresponding to the input address value. In this section, 1 KB SRAM is required to be designed. Hence it requires 10 address bits to cover entire 1 KB memory area. A 5:32 NAND based decoder is designed as row decoder to access 32 bytes of memory area and another 5:32 decoder is used as column decoder in order to access 32 such 32 bytes of memory areas. There by achieving the desired access to 1 KB memory. The schematic of NAND based 5:32 decoder is shown in Fig. 1.7.

The 5:32 NAND based decoder is designed and simulated using LTspice [3]. The simulation results for 5:32 NAND based decoder are shown in Fig. 1.8.

#### 1.2.1.2   Sense Amplifier

A Sense amplifier is an essential circuit in memory chips to speed up the *Read* operation. Due to large arrays of SRAM cells, the resulting signal in the event of *Read* operation has a much lower voltage swing [4]. To compensate for that swing, a sense amplifier is used to amplify voltage coming off Bit Line and ~Bit Line. The voltage coming out of sense amplifier has a full swing voltage of (0–1.8 V). Sense Amplifier also helps reduce the delay times and power dissipation in the overall

**Fig. 1.7** Schematic of NAND based 5:32 row/ column decoder

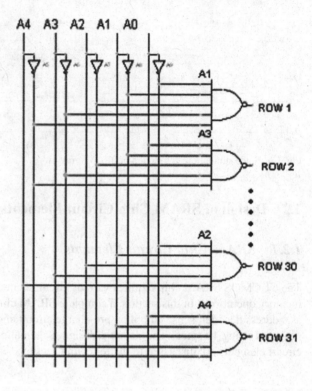

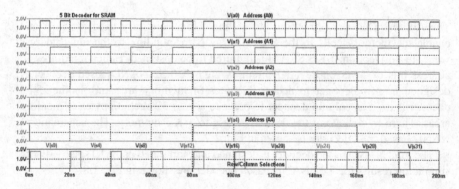

**Fig. 1.8** Spice simulations of 5:32 NAND based row/column decoder

SRAM chip. There are many versions of sense amplifier used in memory chips. The one that is designed in this chapter is a Cross-coupled Sense Amplifier. The schematic of the same is shown in Fig. 1.9.

The Cross-coupled/Feedback Sense amplifier is designed and simulated using LTspice. The LTspice simulations for the same are shown in Fig. 1.10.

**Fig. 1.9** Schematic of
cross-coupled sense amplifier

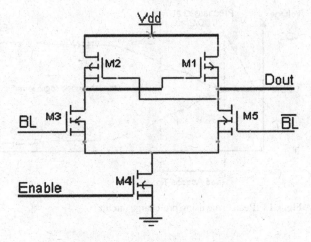

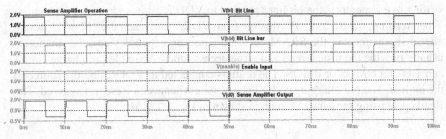

**Fig. 1.10** Spice simulations of cross-coupled sense amplifier

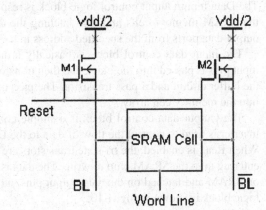

**Fig. 1.11** Schematic of a
pre-charge circuit

### 1.2.1.3   Pre-Charge Circuit

Safe read and write operations require a modification of the memory array and timing sequence based on a pre-charge circuit [5]. The schematic of a pre-charge circuit is shown in Fig. 1.11. The usual voltage of pre-charge is VDD/2. Before reading or

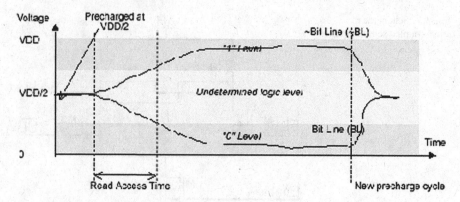

**Fig. 1.12** Read cycle using pre-charge circuit

writing to the memory, the bit lines are tied to VDD/2 using appropriate pass gates. When reading, the BL and ~BL diverge from VDD/2 and reach the "1" or "0" levels after a short time. As the SRAM cells are based on active devices, the memories usually provide the fastest read and write access times. A simple pre-charge circuit consists of a NMOS or PMOS. The drain is connected to VDD/2 and the source to the bit line. The pre-charging on bit lines is done whenever a Reset is triggered. The Read cycle using pre-charge circuit is shown in Fig. 1.12 [5].

### 1.2.1.4    Data I/O Control Logic

The Data Input/Output control logic block is responsible for latching Input data to the SRAM memory cells and also latching the data that needs to be read on the output data ports from the specified address in the SRAM.

The Input data control block is basically a data routing block. Data from the input pins is passed into the block and then transferred to the memory cell array via the buffer circuit and a pass transistor. The pass transistor controls the flow of data into the memory cell array.

The Output data control block is a simple controlled buffer circuit. A tri-state inverter is used to control the flow of data to the Data Out pins on the SRAM chip. When Read is enabled, the tri-state transistors are turned off and prevent data from entering in to the SRAM chip to write. The data is accessed at the specified address on SRAM and latched on the data output pins via Sense amplifier. The I/O control logic block is shown in Fig. 1.13.

## 1.2.2   Design of Complete SRAM Chip

An SRAM chip with 1 KB memory can be built using 32 blocks of 32 bytes array. The design of circuit elements required to support the operation of SRAM chip is

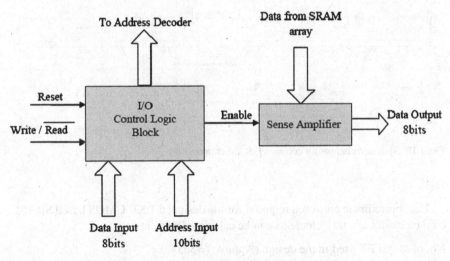

Fig. 1.13   Block diagram of I/O control logic block for 1 KB SRAM chip

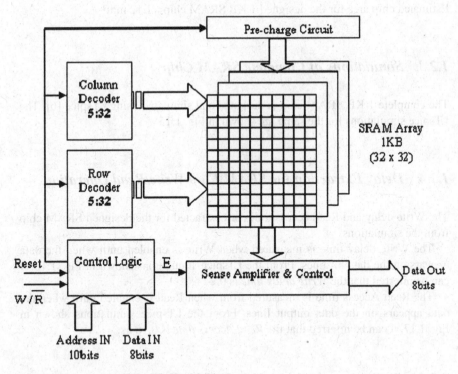

Fig. 1.14   Complete plan of 1 KB SRAM chip

discussed in Sect. 1.2.1. Using these memory cell arrays and circuit elements, a complete 1 KB CMOS based SRAM chip can be designed. In this section, 6T CMOS 1 KB SRAM chip is designed as per the plan shown in Fig. 1.14.

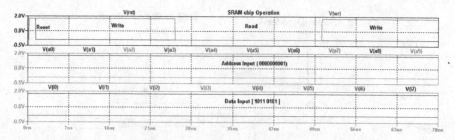

**Fig. 1.15**  Spice simulations for complete SRAM chip operation

The approximate chip area required for the designed 1 KB CMOS based SRAM chip including circuitry elements can be calculated as follows:

No. of MOSFET used in the design (Approx.) = 56,000
Area of a single MOSFET [NMOS/PMOS – average] (Approx.) = 20 $\mu^2$
Total area = 56,000 × 20 $\mu^2$ = 1,120,000 $\mu^2$
Estimated chip area for the designed 1 KB SRAM chip = 1.12 mm²

### 1.2.3   Simulations of Complete SRAM Chip

The complete 1 KB SRAM chip is designed and simulated using LTspice [6]. The LTspice simulations for the same are shown in Fig. 1.15.

### 1.2.4   Delay Extraction for SRAM Chip Write/Read Operation

The Write delay and Read access times are extracted for the designed SRAM chip from the simulations.

The Write delay time is measured when Write is enabled until when the data appears on the data bit lines. From the LTspice simulations shown in Fig. 1.16, it can be inferred that the *Write delay time* is 0.24 ns.

The Read Access time is measured from when Read is enabled until when the data appears on the data output lines. From the LTspice simulations shown in Fig. 1.17, it can be inferred that the *Read Access time* is 0.16 ns.

### 1.2.5   Re-Design of SRAM Chip for Low Power Consumption

The power consumption is very important factor that needs to be considered while designing a chip. It is evident that the SRAM chip is operational whenever the

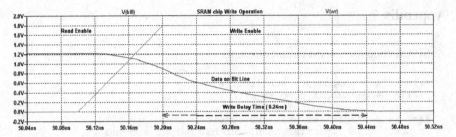

**Fig. 1.16** Spice simulation of SRAM chip to measure write delay time

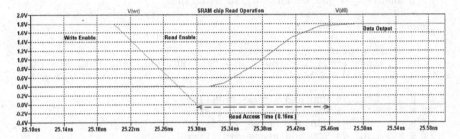

**Fig. 1.17** Spice simulation of SRAM chip to measure read access time

word-line is asserted for read/write operation. The current passes through the cell during read/write operation as long as the word-line is asserted. Hence the power consumption in the chip is directly proportional to the time during which the word-line is asserted.

Based on the above hypothesis, certain measures can be taken by implementing appropriate logic to optimize the power consumption. One of the approaches to the solution for the above mentioned problem is to incorporate clock based assertion of word-line. The word-line may be asserted only for a short and optimized duration for which the write or read operation can be performed completely. Hence the power consumption can be reduced to a certain extent.

The pre-charge voltage that is applied on the bit lines also can be optimized to minimize the power consumption. The duration for which the charge applied on the bit lines may be optimized so that it is just sufficient enough for the sense amplifiers to sense the voltage levels at the shortest time.

Various other measures may be taken based on the floor plan of the transistor, layout, dimensions of transistors, and other factors etc. to optimize power consumption. Additional circuitry also may be incorporated to obtain an optimized and lowest power consuming SRAM chips.

# Appendix

## A. SPICE code for SRAM circuit elements

```
* Interface components subcircuits
* File Name: subckt_comp.sp

**** Sense Amplifier unit subcircuit
.subckt sense_amplifier Enable BL bBL output src gnd

M1 output A src src cmosp l=0.18u w=0.72u
M2 A output src src cmosp l=0.18u w=0.72u

M3 A BL B gnd cmosn l=0.18u w=0.36u
M4 output bBL B gnd cmosn l=0.18u w=0.36u
M5 B Enable gnd gnd cmosn l=0.18u w=0.36u

.ends

**** Sense Amplifier row subcircuit
.subckt sense_amplifier_row E BL0 bBL0 BL1 bBL1 BL2 bBL2 BL3 bBL3
BL4 bBL4 BL5 bBL5 BL6 bBL6 BL7 bBL7 D0 D1 D2 D3 D4 D5 D6 D7 VDD gnd

X0 E BL0 bBL0 N0 VDD gnd sense_amplifier
X1 E BL1 bBL1 N1 VDD gnd sense_amplifier
X2 E BL2 bBL2 N2 VDD gnd sense_amplifier
X3 E BL3 bBL3 N3 VDD gnd sense_amplifier
X4 E BL4 bBL4 N4 VDD gnd sense_amplifier
X5 E BL5 bBL5 N5 VDD gnd sense_amplifier
X6 E BL6 bBL6 N6 VDD gnd sense_amplifier
X7 E BL7 bBL7 N7 VDD gnd sense_amplifier

X8  N0 D0 VDD gnd buffer
X9  N1 D1 VDD gnd buffer
X10 N2 D2 VDD gnd buffer
X11 N3 D3 VDD gnd buffer
X12 N4 D4 VDD gnd buffer
X13 N5 D5 VDD gnd buffer
X14 N6 D6 VDD gnd buffer
X15 N7 D7 VDD gnd buffer

X16 E Dis VDD gnd inverter

M1 D0 Dis gnd gnd cmosn l=0.18u w=0.36u
M2 D1 Dis gnd gnd cmosn l=0.18u w=0.36u
M3 D2 Dis gnd gnd cmosn l=0.18u w=0.36u
M4 D3 Dis gnd gnd cmosn l=0.18u w=0.36u
M5 D4 Dis gnd gnd cmosn l=0.18u w=0.36u
M6 D5 Dis gnd gnd cmosn l=0.18u w=0.36u
M7 D6 Dis gnd gnd cmosn l=0.18u w=0.36u
M8 D7 Dis gnd gnd cmosn l=0.18u w=0.36u

.ends
```

```
**** 5 bit Decoder subcircuit
.subckt decoder_5bit A0 A1 A2 A3 A4 s0 s1 s2 s3 s4 s5 s6 s7 s8 s9
s10 s11 s12 s13 s14 s15 s16 s17 s18 s19 s20 s21 s22 s23 s24 s25 s26
s27 s28 s29 s30 s31 VDD gnd

 X0 A0 bA0 VDD gnd inverter
 X1 A1 bA1 VDD gnd inverter
 X2 A2 bA2 VDD gnd inverter
 X3 A3 bA3 VDD gnd inverter
 X4 A4 bA4 VDD gnd inverter

 X5  bA4 bA3 bA2 bA1 bA0 s0  VDD gnd and5
 X6  bA4 bA3 bA2 bA1  A0 s1  VDD gnd and5
 X7  bA4 bA3 bA2  A1 bA0 s2  VDD gnd and5
 X8  bA4 bA3 bA2  A1  A0 s3  VDD gnd and5
 X9  bA4 bA3  A2 bA1 bA0 s4  VDD gnd and5
 X10 bA4 bA3  A2 bA1  A0 s5  VDD gnd and5
 X11 bA4 bA3  A2  A1 bA0 s6  VDD gnd and5
 X12 bA4 bA3  A2  A1  A0 s7  VDD gnd and5
 X13 bA4  A3 bA2 bA1 bA0 s8  VDD gnd and5
 X14 bA4  A3 bA2 bA1  A0 s9  VDD gnd and5
 X15 bA4  A3 bA2  A1 bA0 s10 VDD gnd and5
 X16 bA4  A3 bA2  A1  A0 s11 VDD gnd and5
 X17 bA4  A3  A2 bA1 bA0 s12 VDD gnd and5
 X18 bA4  A3  A2 bA1  A0 s13 VDD gnd and5
 X19 bA4  A3  A2  A1 bA0 s14 VDD gnd and5
 X20 bA4  A3  A2  A1  A0 s15 VDD gnd and5
 X21  A4 bA3 bA2 bA1 bA0 s16 VDD gnd and5
 X22  A4 bA3 bA2 bA1  A0 s17 VDD gnd and5
 X23  A4 bA3 bA2  A1 bA0 s18 VDD gnd and5
 X24  A4 bA3 bA2  A1  A0 s19 VDD gnd and5
 X25  A4 bA3  A2 bA1 bA0 s20 VDD gnd and5
 X26  A4 bA3  A2 bA1  A0 s21 VDD gnd and5
 X27  A4 bA3  A2  A1 bA0 s22 VDD gnd and5
 X28  A4 bA3  A2  A1  A0 s23 VDD gnd and5
 X29  A4  A3 bA2 bA1 bA0 s24 VDD gnd and5
 X30  A4  A3 bA2 bA1  A0 s25 VDD gnd and5
 X31  A4  A3 bA2  A1 bA0 s26 VDD gnd and5
 X32  A4  A3 bA2  A1  A0 s27 VDD gnd and5
 X33  A4  A3  A2 bA1 bA0 s28 VDD gnd and5
 X34  A4  A3  A2 bA1  A0 s29 VDD gnd and5
 X35  A4  A3  A2  A1 bA0 s30 VDD gnd and5
 X36  A4  A3  A2  A1  A0 s31 VDD gnd and5

 .ends
```

## B. SPICE code for RC parasitic extraction from SRAM cell layout

```
* LINUX                 Sat Mar 24 15:11:27 2007
* PROGRAM advgen
* SPICE LIBRARY

.SUBCKT sram_cell VDD GND BL WL bBL
* Caps2d version: 8
* TRANSISTOR CARDS


MN0     net5#4   WL#3 BL#1      GND#1   nmos    L=0.18U W=1U + effW=1e-06
  MI3/MN0 net5#2   net12#2      GND#3   GND#1   nmos    L=0.18U W=1U      +
effW=1e-06
  MI2/MN0 net12#7 net5#7        GND#4   GND#1   nmos    L=0.18U W=1U      +
effW=1e-06
MN1     net12#4 WL#6 bBL#1      GND#1   nmos    L=0.18U W=1U + effW=1e-06
  MI3/MP0 net5#5   net12        VDD#2   VDD#1   pmos    L=0.18U W=1U      +
effW=1e-06
  MI2/MP0 net12#6 net5#6        VDD#5   VDD#4   pmos    L=0.18U W=1U      +
effW=1e-06


*
* RESISTOR AND CAP/DIODE CARDS
*

Rg1   WL#3    WL#2       45.2083    $poly
Rg2   net12   net12#2   116.2500    $poly
Rg3   net12   net12#3    53.9583    $poly
Rg4   net5#6  net5#7    115.4167    $poly
Rg5   net5#7  net5       35.6250    $poly
Rg6   WL#6    WL#5       46.2500    $poly
Rf1   BL#1    BL          1.0564    $mt1
Rf2   WL#1    WL#2        2.4110    $mt1
Rf3   GND#1   GND#2       2.4778    $mt1
Rf4   GND#1   GND#3       0.2601    $mt1
Rf5   VDD#1   VDD#2       0.3244    $mt1
Rf6   VDD#2   VDD#3       2.4814    $mt1
Rf7   net5    net5#2      1.2832    $mt1
Rf8   net5#2  net5#3      0.3333    $mt1
Rf9   net5#3  net5#4      1.0512    $mt1
Rf10  net5#3  net5#5      0.3238    $mt1
Rf11  VDD#4   VDD#5       0.1669    $mt1
Rf12  VDD#5   VDD#6       2.5082    $mt1
Rf13  GND#1   GND#4       0.3203    $mt1
Rf14  GND#1   GND#5       2.4776    $mt1
Rf15  net12#4 net12#5     1.3201    $mt1
Rf16  net12#5 net12#6     0.3653    $mt1
Rf17  net12#6 net12#3     1.4158    $mt1
Rf18  net12#5 net12#7     0.2855    $mt1
Rf19  WL#4    WL#5        2.4070    $mt1
Rf20  bBL#1   bBL         0.9190    $mt1
Re1   VDD#6   VDD         0.2832    $mt2
Re2   VDD     VDD#3       1.2223    $mt2
Re3   GND#5   GND         0.7362    $mt2
Re4   GND     GND#2       1.3788    $mt2
Re5   WL#4    WL          2.2416    $mt2
Re6   WL      WL#1        2.0018    $mt2
```

```
*
*         CAPACITOR CARDS
*

C1    VDD     GND    2.038E-16
C2    BL      GND    2.661E-16
C3    WL      GND    6.774E-16
C4    bBL     GND    2.865E-16
C5    net12   GND    1.597E-16
C6    net5    GND    4.261E-16
C7    net5#6  GND    2.298E-16
C8    WL#6    GND    7.296E-17
C9    net5#7  GND    1.507E-16
C10   net12#2 GND    2.056E-16
C11   WL#3    GND    7.003E-17
C12   WL#5    GND    3.171E-16
C13   net12#3 GND    4.754E-16
C14   WL#2    GND    3.062E-16
C15   WL#4    GND    7.411E-16
C16   VDD#6   GND    4.331E-16
C17   VDD#3   GND    6.310E-16
C18   WL#1    GND    6.933E-16
C19   VDD#4   GND    1.635E-16
C20   bBL#1   GND    7.459E-16
C21   net12#4 GND    2.806E-16
C22   net12#7 GND    1.575E-16
C23   net5#2  GND    3.270E-16
C24   net5#4  GND    2.873E-16
C25   BL#1    GND    7.836E-16
C26   VDD#5   GND    1.549E-16
C27   net12#6 GND    3.979E-16
C28   net5#5  GND    1.381E-16
C29   VDD#2   GND    1.350E-16
C30   VDD#1   GND    1.746E-16
C31   net5#3  GND    3.524E-16
C32   net12#5 GND    3.510E-16

.ENDS sram_cell
```

# References

1. Kang S, Leblebici Y (2003) CMOS digital integrated circuits, 3rd edn. Tata McGraw-Hill, Boston
2. Static Random Access Memory Interface (2007) EE Herald. http://www.eeherald.com/section/design-guide/esmod15.html. Accessed 4 June 2007
3. LT Spice User Guide (2006) Linear technology. http://LTspice.linear-tech.com/software/scad3.pdf. Accessed 10 August 2006
4. Mehata K, Zinkovski I (2002) CSE447: Design of 1 KB SRAM chip. The Pennsylvania State University. http://www.cedcc.psu.edu/khanjan/vlsihome.htm. Accessed 4 June 2007
5. Static RAM Memory (2006) Institut National des Sciences. https://intranet.insa-toulouse.fr/view/422/content/static_ram.html. Accessed 10 August 2006
6. ECE558: Spice Simulations (2006) University of Massachusetts http://www-unix.ecs.umass.edu/~zzeng/ece558/spice_www/spice.html. Accessed 10 August 2006

# Chapter 2
# FPGA Application Design

In wired or wireless communication systems, the information that needs to be transmitted is not only required to reach the destination but it should be error free and should make efficient use of the channel bandwidth available. Various DSP based encoding/decoding algorithms, data compression and noise filtering techniques have been developed to achieve effective and efficient data transmission with the help of FPGAs for hardware implementation. FPGA based implementations provide the flexibility of re-programming and quick delivery of the product to the market.

This chapter demonstrates the design of a simple DS-SS system including the basic building blocks such as, PN sequence generator, BPSK modulator/demodulator, BOOTH multiplier, Low Pass Filter and convolutional coding. The system is designed using Verilog HDL, simulation and functional verification of the design is performed using ModelSim® XE III 6.0d, and synthesis using Xilinx® ISE. The design is implemented and tested on Xilinx® Spartan 2E FPGA.

This chapter also demonstrates some of the algorithms and techniques used to accomplish data integrity and channel bandwidth efficiency in a communication system such as, Low Pass FIR filter using efficient Distributed Arithmetic (DA) architecture, Discrete Cosine Transform (DCT) using Scaled DCT architecture and Convolution coding and Viterbi decoding techniques. The Low Pass-Finite Impulse Response (LP-FIR) filter coefficients are calculated using MatLab FDA tool based on the given specification of the filter. The systems are designed using Verilog HDL, simulation and functional verification of the design is done using ModelSim® XE II 6.0d and synthesis using Xilinx® ISE. The designs are implemented on Xilinx® Spartan 2E FPGA.

The prerequisites for approaching this chapter would be an adequate background of basic digital communication system.

V.A. Chandrasetty, *VLSI Design: A Practical Guide for FPGA and ASIC Implementations*, SpringerBriefs in Electrical and Computer Engineering, DOI 10.1007/978-1-4614-1120-8_2, © Springer Science+Business Media, LLC 2011

## 2.1 Design of Direct Sequence-Spread Spectrum System

Direct Sequence-Spread Spectrum (DS-SS) is a transmission technique in which a pseudo-noise code, independent of the information data is employed as a modulation waveform to "spread" the signal energy over a bandwidth much greater than the signal information bandwidth. At the receiver the signal is *de-spread* using a synchronized replica of the pseudo-noise code. The spreading sequence in DS-SS is often called as PN sequence.

In this section, the spread signal is modulated using Binary Phase Shift keying (BPSK) modulation technique in the transmitter and on the receiver side the modulated signal is recovered using BPSK demodulation technique.

The basic building blocks of DS-SS system are shown in Fig. 2.1 [1].

### 2.1.1 PN Sequence Generator

#### 2.1.1.1 Overview of PN Sequence Generator

A Pseudo-random Noise (PN) sequence/code is a binary sequence that exhibits randomness properties but has a finite length and is therefore deterministic. PN generators are heart of every spread spectrum systems. Each symbol or bit in the sequence is called as *Chip* [2].

PN generators are based on Linear Feedback Shift Registers (LFSR). The contents of the registers are shifted right by one position at each clock cycle. The feedback from predetermined registers or taps to the left most register are XNOR-ed together.

LFSRs have several variables:

- The number of stages in the shift registers
- The number of taps in the feedback path
- The position of each tap in the shift registers stage
- The initial starting condition of the shift register often referred to as the "FILL" state

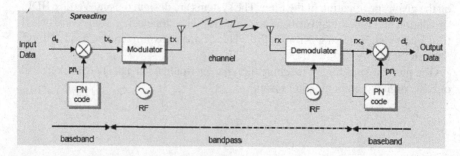

**Fig. 2.1** Basic building blocks of DS-SS system

The longer the number of stages of shift registers in the PN generator, longer the duration of the PN sequence before it repeats. For a shift register of fixed length N, the number and duration of the sequences that it can generate are determined by the number and position of taps used to generate the parity feedback bit.

A maximum length sequence (L) for a shift register of length N is referred to as m-sequence and is defined as [3]:

$$L = 2^N - 1,$$

E.g. an eight stage LFSR will have a set of m-sequences of length 255.

Some of the most popular types of PN Sequence generators are:

- m-sequence codes
- Barker codes
- Gold codes

### 2.1.1.2   Design of PN Sequence Generator

Design

*Specifications:*

- Clock frequency for PN sequence generator system, $F_{pn} = 100$ KHz.
- LFSR length, $N = 4$.
- LFSRs are of D-FF type.
- X-NOR gate is used for linear parity feedback to the system.
- FPGA board clock frequency, $F_b = 50$ MHz (assumption)

*Procedure:*

- A clock frequency of 100 KHz for PN Sequence generator is designed using a divider of 500 clock cycles of $F_b$.
  Clock divider $= F_b/F_{pn} = 50$ MHz/100 KHz $= 500$
- Maximum length sequence, $N = 4$ corresponds to 4 D-FF to realize LFSRs of the PN generator system.
  Since $N = 4$, the maximum sequence length $L = 2^4 - 1 = 15$.
  Hence the sequence repeats every 15 clock cycles.
- The Chip rate for the PN sequence generator system is calculated as follows:
  Chip period, $T_c = 1/100$ KHz $= 10$ μs
  Chip rate, $F_c = 100$ KHz
- The bit period for the input data signal is calculated as follows:
  Data bit period, $T_d = $ Max. sequence Length (L) × Chip period (Tc)
  For the system, $T_d = 15 \times 10$ μs
  Hence, the input data bit period for the system is, $T_d \geq 150$ μs.

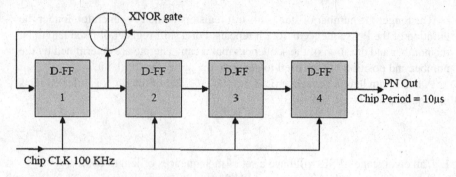

**Fig. 2.2**  Block diagram of a PN sequence generator

Block Diagram

The block diagram of a PN sequence generator for the design specification is shown in Fig. 2.2.

### 2.1.1.3   Properties of PN Sequence

*Merits of using PN sequence* [4]:

1. *Balance property*: In each period of the sequence the number of binary ones differ from the number of binary zeros by at most one digit (when LFSR stage length is odd)

$$Pn = +1+1+1-1-1+1+1-1 = +1$$

2. *Run-length Distribution*: A run is a sequence of a single type of binary digits. Among the sequence of ones and zeros in each period it is desirable that one-half the runs of each type are of length 1, about one-fourth are of length 2, one-eight are of length 3 and so on.
3. *Autocorrelation*: The origin of the name pseudo-noise is that the digital signal has an autocorrelation function which is very similar to that of a white noise signal. For PN sequences the autocorrelation has a large peaked maximum for perfect synchronization of two identical sequences (like white noise). The synchronization of receiver is based on this property.
4. *Cross-correlation*: Cross-correlation is the measure of agreement between two different codes $pn_1$ and $pn_2$. When Cross-correlation is zero the codes are called Orthogonal. In CDMA multiple users occupy the same RF bandwidth and transmit simultaneously. When the user codes are orthogonal, there is no

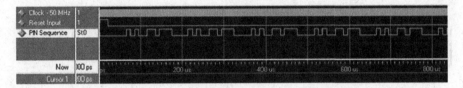

**Fig. 2.3** Simulation results for PN sequence generator

interference between the users after dispreading and the privacy of the communication of each user is protected.

*Demerits of using PN sequence* [4]:

1. *Synchronization*: The most sensitive aspect of DS-SS system is the synchronization of the transmitter's PN sequence to that of the receiver where an offset of even one PN chip can result in noise rather than a de-spread symbol sequence.
2. *Increased Bandwidth*: As the data signal is spread using PN codes at higher frequency, there is an increase in bandwidth used in the process.
3. *Complexity*: There is an increased complexity and computational load both in the receiver and the transmitter to spread/de-spread the signal.

#### 2.1.1.4 Simulation Results for PN Sequence Generator

The PN sequence generator is designed using Verilog HDL. Functional verification and simulation is performed using ModelSim.

The simulation results for PN sequence generator is shown in Fig. 2.3.

### 2.1.2 Transmitter for Direct Sequence-Spread Spectrum System

#### 2.1.2.1 Overview of DS-SS Transmitter System

In DS-SS transmitter, the input data bits are spread by PN sequence generator. The spreading is actually done by multiplying the data bits with that of the PN sequence code generated. The frequency of PN sequence is higher than the Data signal. After spreading, the Data signal is modulated and transmitted. There are several schemes available for modulation, viz. BPSK, QPSK, M-QAM etc. The most widely used modulation scheme is the BPSK. In this design, BPSK modulation is used to modulate and transmit the spread signal.

The basic building blocks of a simple DS-SS transmitter system are shown in Fig. 2.4.

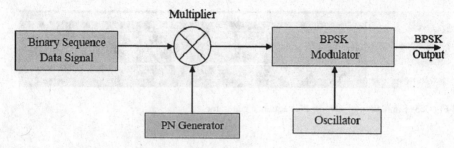

**Fig. 2.4** Block diagram of a DS-SS transmitter system

**Table 2.1** Truth table for the multiplier

| m(t) | p(t) | s(t) |
|------|------|------|
| 0 | 0 | 1 |
| 0 | 1 | 0 |
| 1 | 0 | 0 |
| 1 | 1 | 1 |

### 2.1.2.2  Design of DS-SS Transmitter

Multiplier Design

*Specifications*:

- PN sequence Chip rate, $Tc = 10$ μs.
- Data signal Bit rate, $Tb \geq 150$ μs.

Let the data signal be m(t) and PN sequence p(t). The two signals are multiplied and the multiplied output is the spread signal. Truth table for the multiplier $s(t) = m(t).$ p(t) is shown in Table 2.1.

From the truth table, it can be inferred that an XNOR gate can act as a multiplier to spread the data signal with the PN signal. Hence the block diagram for the multiplier is shown in Fig. 2.5.

Oscillator Design

*Specification*:

- PN sequence Chip rate, $Tc = 10$ μs.
- Carrier frequency, $Fc \geq 5$ times Chip rate.

*Design*:

- The oscillator carrier sampling rate is designed
  Let the Sampling rate of sine wave be $Fs = 25$ MHz.

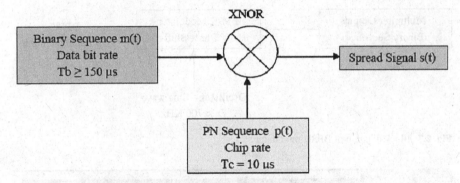

**Fig. 2.5**  Block diagram of a data and PN sequence multiplier

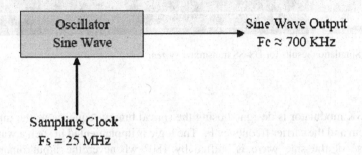

**Fig. 2.6**  Block diagram of an oscillator

- Number of samples for a full cycle of sine wave is designed
    Let the number of samples for a full cycle be $N = 36$.
- The oscillator is designed to generate sine wave of carrier frequency Fc
    $Fc \geq 5(1/T_C) = 5(1/10\mu s) = 500 KHz$.

For the above design with sampling rate 25 MHz and 36 samples per cycle, the carrier frequency, $Fc = 25$ MHz/$36 \approx 700$ KHz. The oscillator is implemented using a Look-Up-Table (LUT) of nine samples and the logic is design in order to oscillate generating a sine wave.

The block diagram of the oscillator as per the design is shown in Fig. 2.6.

## BPSK Modulator Design

*Specification:*

- Spread binary sequence is the input to the system
- Oscillator carrier sine wave of frequency, $Fc \approx 700$ KHz

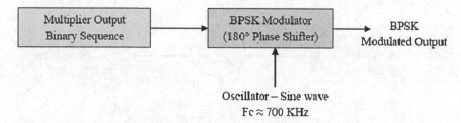

**Fig. 2.7**   Block diagram of BPSK modulator

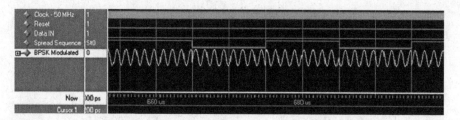

**Fig. 2.8**   Simulation results for DS-SS transmitter system

*Design*:

The BPSK modulator is designed using the spread binary sequence as the input to the system and the carrier frequency $F_c$. The logic is implemented in such a way that the phase of the sine wave is shifted by 180° whenever the input binary bit changes.

The block diagram of the BPSK Modulator as per the design is shown in Fig. 2.7.

### 2.1.2.3   Simulation Results for DS-SS Transmitter

The DS-SS transmitter is designed using Verilog HDL. Functional verification and simulation is done using ModelSim. The simulation results for DS-SS transmitter is shown in Fig. 2.8.

## 2.1.3   Receiver for Direct Sequence-Spread Spectrum System

### 2.1.3.1   Overview of DS-SS Receiver System

In DS-SS receiver, the input to the system is the BPSK modulated signal. This signal would have been affected by noise and other interference in the communication channel. The DS-SS receiver should be designed carefully to reproduce the data signal with least error.

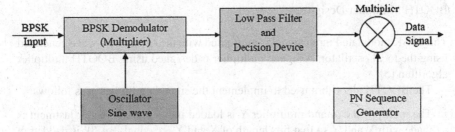

**Fig. 2.9**  Block diagram of a DS-SS receiver system

The BPSK modulated input signal is multiplied by the locally generated carrier wave by the oscillator. The multiplied signal is then passed through the low pass filter to get low frequency components only. A decision device is used to approximate the signal to binary sequence. This binary sequence is the spread sequence of the data signal.

The most sensitive part of the DS-SS receiver is the synchronization of the locally generated PN sequence and the sequence obtained from the decision device [3]. Even a single bit mismatch may lead to noise instead of the data signal. Suitable technique is used to achieve synchronization and multiply the local PN sequence code with that of the received PN code. The Data signal is obtained after the multiplication process.

In this design, since transmitter and receiver uses common clock on the same FPGA board, the delay in the receiver is considered and modeled appropriately. No specific synchronization technique is used.

The block diagram of a simple DS-SS receiver system is shown in Fig. 2.9.

### 2.1.3.2  Design of DS-SS Receiver

BPSK Demodulator Design

*Specifications*:

- BPSK modulated signal is the input to the system
- Oscillator carrier sine wave of frequency, $Fc \approx 700$ KHz

The input BPSK signal is multiplied with the carrier sine wave generated from the local oscillator. The design and implementation of the signed BOOTH multiplier is discussed in the following section.

The multiplied output will have higher frequency components and channel noise as well. The high frequency components are eliminated using a suitable Low Pass Filter. Design of rectangular window Low-Pass FIR filter is also discussed in the following section.

The filtered low frequency component will have distortion in the signal. Hence a suitable 'Decision Device' is used to smoothen to binary sequence.

BOOTH Multiplier Design

The BPSK modulated input signal is multiplied with the carrier sine wave generated using the local oscillator. A signed multiplier is designed using BOOTH multiplier algorithm [5].

The BOOTH algorithm used to implement the signed multiplier is as follows:

- The multiplicand X and multiplier Y is loaded into a register. Bit adjustment is made with X and Y so that bits length of X and Y are equal. Bit '0' is padded in order to achieve it
- An accumulator is used to store the result. The length of the accumulator should be twice the length of multiplicand or multiplier. A = 2X or 2Y
- The multiplicand X is loaded into the accumulator from LSB
- A dummy bit of 0 is appended with the accumulator A at the LSB
- During the multiplication operation, the pair of LSB of the accumulator and the dummy bit is considered to follow further arithmetic operations
- Depending on the bit pair obtained in the previous step, following operations are performed:

  o "00" – Arithmetic shift right of the Accumulator.
  o "01" – Add multiplier Y to the Accumulator A (from MSB of A) and Arithmetic shift right of Accumulator.
  o "10" – Subtract multiplier Y from the Accumulator A (from MSB of A) and Arithmetic shift right of Accumulator.
  o "11" – Arithmetic shift right of the Accumulator.

  Shift operations are performed along with dummy bit.
- The above operations are continued till MSB of multiplicand X is shifted off from the accumulator A.

In this section, 5-bit signed BOOTH multiplier is designed and implemented.

Low Pass Filter and Decision Device Design

*Specifications*:

- The multiplied output from the BPSK demodulator is the input to this system
- A Low Pass Filter with cutoff frequency, f = 105 KHz
- Oscillator carrier wave sampling rate, Fs = 25 MHz

*Design*:

A Rectangular window FIR filter is designed with a cutoff frequency, f = 105 KHz. Let the length of impulse response for the filter, N = 2.
The desired response of the ideal Low-pass filter is given by,

$$H_d(e^{jw}) = 1, 0 \leq f \leq 105 \text{ KHz, otherwise } 0$$

The normalized angular frequency, $\omega_c = 2\pi F/Fs = 8.4\pi \times 10^{-3}$

$$H_d(e^{jw}) = 1, 0 \leq \omega \leq \omega_c; \ 0, \omega_c \leq \omega \leq \pi$$

The filter coefficients are given by,

$$h_d(n) = \sin(8.4\pi \times 10^{-3} \text{ N}) / (\pi \text{N}), \text{ where N } \neq 0.$$

Therefore, the filter coefficients are,

$$h(0) = 8.40 \times 10^{-3} \text{ and } h(1) = 8.39 \times 10^{-3}$$

In this design, one sample of the signal is stored in a register and then it's added with the next sample. The filtered output samples obtained is then processed by the Decision Device. The output of the Decision Device is held High (1) when the output of the filter is non-negative otherwise it's made Low (0).

### 2.1.3.3 Noise Models and Synchronization

*Noise models* [1]*:*

- *Multi Path Channels*: In wireless channels there exists often multi path propagation. Since there are more than one path from the transmitter to the receiver. Such multi paths may be due to (a) atmospheric reflection or refraction (b) Reflections from ground, buildings or other objects. Corrective actions are taken to eliminate noise due to multi path channels using appropriate synchronization techniques.
- *Jamming*: The goal of the jammer is to disturb the communication of his adversary. Protection against jamming waveforms is provided by purposely making the information-beating signal occupy a bandwidth far in excess of the minimum bandwidth necessary to transmit it. This has the effect of making the transmitted signal assume a noise-like appearance so as to blend into background. The transmitted signal thus enabled to propagate through the channel undetected by anyone who may be listening. Spread spectrum is a method of "camouflaging" the information bearing signal.

In this design, the noise effect is not modeled as the transmitter and receiver is on the same FPGA board without any air interface.

*Synchronization techniques* [1]*:*

For proper operation of DS-SS system, the locally generated PN sequence in the receiver is synchronized to the PN sequence of the transmitter generator in both its

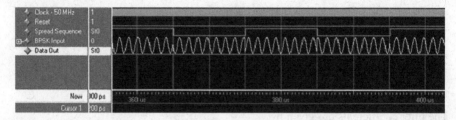

**Fig. 2.10**  Simulation results for DS-SS receiver system

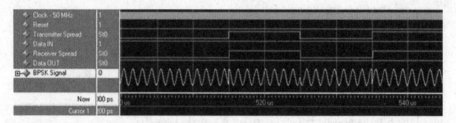

**Fig. 2.11**  Simulation results for DS-SS modem

rate and position. A slight misalignment in the sequence results in noise instead of data signal.

The process of synchronizing the locally generated PN sequence with the received PN sequence is usually accomplished in two steps. The first step called *acquisition* consists of bringing the two spreading signals into coarse alignment with one another. Once the received PN sequence has been acquired, the second step called *tracking* takes over and continuously maintains the best possible waveform fine alignment by means of a feedback loop. This is essential to achieve highest correlation power and thus highest processing gain (SNR) at the receiver.

In this design, synchronization technique is not modeled since the same clock and PN sequence for receiver and transmitter is implemented on the same FPGA board. A delay of one clock pulse is modeled while multiplying the PN code in the receiver to compensate the filtering delay of one sample.

### 2.1.3.4  Simulation Results for DS-SS Receiver

The DS-SS receiver is designed using Verilog HDL [6]. Functional verification and simulation is done using ModelSim.

The simulation results for DS-SS receiver is shown in Fig. 2.10.

The simulation results for DS-SS modem is shown in Fig. 2.11. The synthesis report obtained from Xilinx ISE is also shown in Fig. 2.12. The modem can operate at a maximum frequency of 64 MHz on Xilinx Spartan 2E FPGA.

```
====================================================================
HDL Synthesis Report

Macro Statistics
 # Adders/Subtractors                                              : 17
   10-bit adder                                                    : 1
   11-bit adder                                                    : 10
   32-bit adder                                                    : 4
   5-bit adder                                                     : 2
 # Counters                                                        : 5
   32-bit up counter                                               : 3
   32-bit updown counter                                           : 2
 # Registers                                                       : 33
   1-bit register                                                  : 29
   10-bit register                                                 : 2
   5-bit register                                                  : 2
 # Comparators                                                     : 3
   33-bit comparator greatequal                                    : 3
 # Multiplexers                                                    : 5
   11-bit 4-to-1 multiplexer                                       : 5
 # Xors                                                            : 8
   1-bit xor2                                                      : 8

====================================================================
```

**Fig. 2.12** Synthesis report for DS-SS modem

## 2.2   FIR Filter Design

### 2.2.1   Concepts of FIR Filter

A discrete-time filter produces a discrete-time output sequence for the discrete-time input sequence. In the Finite Impulsive Response (FIR) system, the impulse response sequence is of finite duration, i.e. it has a finite number of non-zero terms and hence the filter coefficients are also constant. The response of the FIR filter depends only on the present and past input samples (a causal system). Thus making the system always stable.

The difference equation for length 'M' FIR filter is given by [4],

$$y(n) = b_0 \times (n) + b_1 \times (n-1) + b_2 \times (n-2) + b_3 \times (n-3) + \ldots b_{M-1} \times (n-M+1)$$

$$Y(n) = \sum_{K=0}^{M-1} b_k \times (n-K)$$

where, $[b_k]$ is the set of filter coefficients.

Some of the important characteristics of FIR digital filter are as follows [4]:

• They can have an exact linear phase
• They are always stable
• The design methods are generally linear
• They can be realized efficiently in hardware
• The filter start-up transients have finite duration
• The filter coefficients are constant for the given order of the filter

**Table 2.2** Filter coefficients for LP FIR filter with order 16

| Transfer function | Coefficients | Transfer function | Coefficients |
|---|---|---|---|
| h(0) | 0.0328 | h(8) | 0.5763 |
| h(1) | 0.0816 | h(9) | −0.0550 |
| h(2) | −0.0065 | h(10) | −0.0694 |
| h(3) | −0.0047 | h(11) | 0.0847 |
| h(4) | 0.0847 | h(12) | −0.0047 |
| h(5) | −0.0694 | h(13) | −0.0065 |
| h(6) | −0.0550 | h(14) | 0.0816 |
| h(7) | 0.5763 | h(15) | 0.0328 |

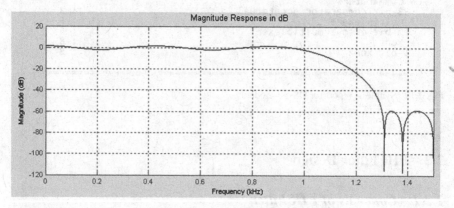

**Fig. 2.13** Frequency response (Magnitude) for the designed LP FIR filter

In this section a Low-Pass FIR filter is designed using MatLab FDA tool for the given specifications. Simulated using ModelSim® and implemented using Xilinx® 2E FPGA.

### 2.2.2  Low Pass FIR Filter Design

The Low Pass FIR (LPF) specifications given in the assignment are,

- $F_{pass} = 1$ KHz, $F_{stop} = 1.3$ KHz
- Pass band ripple = 3 dB, Stop band ripple = 60 dB

Assuming,

- Sampling frequency of the input signal, $F_s = 3$ KHz.
- FIR Filter design method: Equiripple with density factor 16.

The filter coefficients are obtained using MatLab FDA tool for the given specification. The order of the filter, $N = 16$. The filter coefficients h(n) are as shown in Table 2.2. The frequency response for the given filter specification is shown in Fig. 2.13

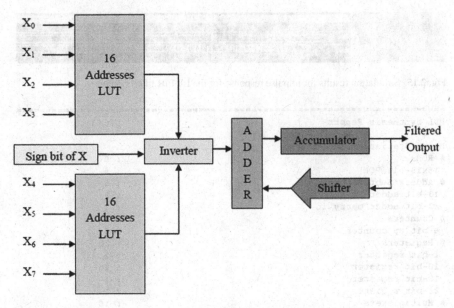

**Fig. 2.14**  Block diagram to illustrate the functional operation of DA architecture

### 2.2.3   Distributed Arithmetic Architecture

Distributed Arithmetic (DA) is an important technique to implement digital signal processing functions in FPGAs. DA provides an approach for multiplier-less implementation of DSP systems. It is an algorithm that can perform multiplication with Look-Up Table (LUT) based schemes. DA specifically targets the sum of products (also referred to as the vector dot product) computation that is found in many of the important DSP filtering and frequency transforming functions [7].

In this section, LP FIR filter is designed and implemented using DA architecture. By observing the filter coefficients in Table 2.2, the second half (8–15) of filter coefficients are mirror image of the first half (0–7). Hence the SOP for second half can be accessed from the first half by re-ordering the input bits appropriately. The first half (0–7) coefficients can be broken into two parts and SOP can be calculated and stored in two different blocks. Hence, two LUTs of length 16 are sufficient to store the SOP for the obtained filter coefficients.

The basic functional operation of DA architecture is shown in Fig. 2.14.

### 2.2.4   Simulation and Synthesis Results

The LP FIR filter is designed using Verilog HDL. The design is simulated using ModelSim®. The impulse response for the LP FIR filter system is shown in Fig. 2.15. In this design, fixed point representations of real numbers are used. Filtered output

**Fig. 2.15** Simulation results for impulse response for the LP FIR filter system

```
===============================================================================
HDL Synthesis Report

Macro Statistics
# ROMs                                                  : 4
 16x13-bit ROM                                          : 4
# Adders/Subtractors                                    : 4
 13-bit adder                                           : 3
 22-bit adder carry in                                  : 1
# Counters                                              : 1
 4-bit up counter                                       : 1
# Registers                                             : 20
 1-bit register                                         : 1
 10-bit register                                        : 16
 13-bit register                                        : 1
 22-bit register                                        : 2
# Multiplexers                                          : 16
 1-bit 10-to-1 multiplexer                              : 16
# Logic shifters                                        : 2
 22-bit shifter logical left                            : 1
 32-bit shifter logical left                            : 1

===============================================================================
```

**Fig. 2.16** HDL synthesis report for LP FIR filter design

values have lower 8 bits representing decimal part. Hence the exact filtered output values from the simulation results are calculated as follows:

$$Y = (8, 22, -2, -12, 22, -18, -13, 148, 148, -13, -18, 22, -12, -2, 22, 8)/2^8$$
$$Y = (0.0312, 0.8593, -0.0078, -0.0468, 0.8593, -0.0703, -0.0507, 0.5781,$$
$$0.5781, -0.0507, -0.0703, 0.8593, -0.0468, -0.0078, 0.8593, 0.0312)$$

The design is synthesized and implemented on Xilinx® Spartan 2E FPGA. The HDL synthesis report is shown in Fig. 2.16.

## 2.3   Discrete Cosine Transform Algorithms

### 2.3.1   Concepts of DCT

The Discrete Cosine Transform (DCT) is a technique that converts a spatial domain waveform into its constituent frequency components as represented by a set of coefficients. The process of reconstructing a set of spatial domain samples is called the Inverse Discrete Cosine Transform (IDCT). The equation for 1-D N-point DCT is given by [8],

$$X(k) = \alpha(k) \sum_{n=0}^{N-1} x(n) \cos[\frac{\pi(2n+1)k}{2N}] \quad 0 \leq k \leq N-1$$

where,

$$\alpha(0) = \sqrt{\frac{1}{N}}, \quad \alpha(k) = \sqrt{\frac{2}{N}} \; for \; 1 \leq k \leq N-1$$

One-Dimensional DCT has most often been used in two-dimensional DCT by employing the row-column decomposition which makes it suitable for hardware implementation. Typically the DCT coefficients produced have most of the block's energy in a few frequency domain elements and hence quantization and coding is applied after DCT to provide lossless as well as lossy actual compression [8].

For data compression of image/video frames, usually a block of data is converted from spatial domain samples to another domain (usually frequency domain) which offers more compact representation. DCT technique is used in a wide range of signal and image processing applications. Some of the most popular applications are [8],

- JPEG and JPEG2000 image compression standards
- MPEG digital video standards
- H.261 and H.263 video conferencing standards
- Progressive Image Transmission (PIT) systems: teleconferencing, medical diagnostic imaging and security services

### 2.3.2 DCT Architectures on FPGA

The DCT can be implemented on FPGA using various architectures. Some of the popular one's reported in [9] are discussed below:

- *Distributed Arithmetic*: The N-points DCT can be considered as N parallel filters. The DCT on the array requires N shift registers for parallel-to-serial conversion, N LUT memories and N shift-accumulators. All the N memories receive the same address. One shift-register and a shift-accumulator are each mapped to an add-shift cluster, while the LUT is mapped to a part of a memory cluster.
  *Area usage*: 8 shift registers + 8 ROMs + 8 Accumulators
- *Mixed ROM*: The 8-point 1D-DCT can be expressed as the product of an 8×8 matrix by an eight element column vector. Through algebraic manipulations, this matrix can be reduced to 4×4 matrix. Hence, the number of words per ROM is reduced to only 16 but some overhead has been incurred in the form of adders to calculate the address of the ROMs.
  *Area usage*: 4 adders + 4 subtractions + 8 shift registers + 8 accumulators + 8 ROMs

- *CORDIC Rotator based:* The DCT computation is done using CORDIC rotator [10]. Since the memory is an integral part of the DA, and ROM size increases exponentially with respect to vector size N. Many techniques have been developed for reducing the size of ROM. The CORDIC algorithm reformulates the 1-D DCT so that the ROM size is reduced to a fix size of four words, independent of the bandwidth of the input data. The DA functionality is implemented by converting parallel data to serial through shift registers and using this data to formulate the address of the memories. This implementation requires 6-CORDIC and 16 butterfly adders for an 8-point 1-D DCT. The CORDIC rotators are implemented through ROM and shift accumulators, while butterfly adders are implemented through add-shift clusters [11].

  *Area usage:* 8 adders + 8 subtractions + 8 shift registers + 12 accumulators + 12 ROMs
- *Skew circular convolution:* This technique starts with re-ordering the input sequences. Then skew circular convolutions are performed on the reordered inputs, which give odd-indexed transformed sequence. The transformed sequences are re-ordered for the proper output sequences.

  *Area usage:* 4 adders + 4 subtractions + 8 shift registers + 8 accumulators + 8 ROMs

### 2.3.3 Scaled 1-D 8-Point DCT Architecture

Since using LUTs results in a very efficient and regular structure suitable for VLSI implementation, especially on the FPGAs, there has been great interest in developing similar kind of LUT based DCT architecture. The Scaled DCT architecture is also a LUT based design. The architecture is primarily designed by making mathematical and trigonometric manipulation using 1-D 8-point DCT equation on eight input data samples. In this design, LUT based Distributed Arithmetic architecture is used. The basic building blocks of this architecture are [9]:

- 20 butterfly adders
- 12 shift registers
- 10 LUTs

The constant scale factor (Y0 and Y4) is not considered in this implementation as that can be combined with the quantization constants without requiring any additional hardware such as LUTs. The simplified 1-D 8-point DCT equations are as shown below:

$$Y_0 = \left[\sqrt{2} \times (X_0 + X_1 + X_2 + X_3 + X_4 + X_5 + X_6 + X_7)\right]/4$$
$$Y_1 = \left[(X_0 - X_7) \times A + (X_1 - X_6) \times B + (X_2 - X_5) \times C + (X_3 - X_4) \times D\right]/2$$
$$Y_2 = \left[(X_0 + X_7 - X_3 - X_4) \times E + (X_1 + X_6 - X_2 - X_5) \times F\right]/2$$

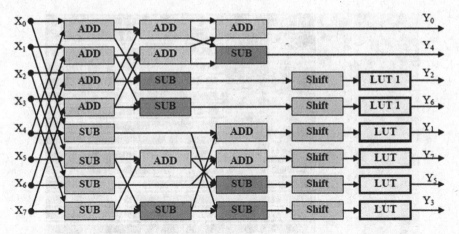

**Fig. 2.17** Block diagram of scaled DCT architecture

$$Y_3 = \left[(X_0 - X_7) \times B + (X_6 - X_1) \times D + (X_5 - X_2) \times A + (X_4 - X_3) \times C\right]/2$$

$$Y_4 = \left[\sqrt{2} \times (X_0 - X_1 - X_2 + X_3 + X_4 - X_5 - X_6 + X_7)\right]/2$$

$$Y_5 = \left[(X_0 - X_7) \times C + (X_6 - X_1) \times A + (X_2 - X_5) \times D + (X_3 - X_4) \times B\right]/2$$

$$Y_6 = \left[(X_0 + X_7 - X_3 - X_4) \times F + (X_2 + X_5 - X_1 - X_6) \times E\right]/2$$

$$Y_7 = \left[(X_0 - X_7) \times D + (X_6 - X_1) \times C + (X_2 - X_5) \times B + (X_4 - X_3) \times A\right]/2$$

For $N = 8$,
$A = \cos(\pi/16)$
$B = \cos(3\pi/16)$
$C = \cos(5\pi/16)$
$D = \cos(7\pi/16)$
$E = \cos(\pi/8)$
$F = \cos(3\pi/8)$

The constant values A, B, C, D, E and F that is required to be multiplied with input X is performed by LUT based Distributed Arithmetic architecture. The block diagram of Scaled DCT architecture for 1-D 8-point samples is shown in Fig. 2.17.

### 2.3.4  Simulation and Synthesis Results

In this section, 1-D 8-point DCT is designed using Scaled DCT architecture and coded in Verilog HDL. The design is simulated using ModelSim®. The DCT for the input samples, $X = (4, 2, 8, 4, 4, 6, 6, 6)$ is as shown in Fig. 2.18.

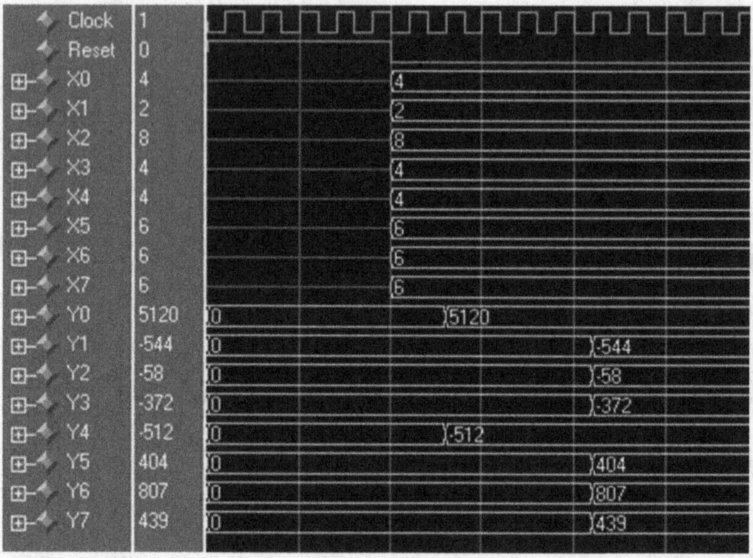

**Fig. 2.18** Simulation results for 1-D 8-point DCT

$$Y = (5120 / \sqrt{2}, -544, -58, -372, -512 / \sqrt{2}, 404, 807, 439)$$

In this design, fixed point representations of real numbers are used. DCT output values have lower eight bits representing decimal part of DCT output. Hence the exact DCT output values from the simulation results are calculated as follows:

$$Y = (5120 / \sqrt{2}, -544, -58, -372, -512 / \sqrt{2}, 404, 807, 439) / 2^8$$

$$Y = (14.1421, -2.0882, -0.2242, -1.4221, -1.4142, 1.6011, 3.1543, 1.7475)$$

This design is implemented on Xilinx® Spartan 2E FPGA. The HDL [13] synthesis report is shown in Fig. 2.19.

## 2.4  Convolution Codes and Viterbi Decoding

### 2.4.1  Concepts of Convolution Codes

Forward Error Correction (FEC) technique is used to improve the capacity of channel by adding some carefully designed redundant information to the data that is transmitted over the communication channel. The process of adding this redundant information is known as *channel coding*.

```
============================================================================
HDL Synthesis Report

Macro Statistics
# ROMs                                                        : 10
 4x16-bit ROM                                                 : 10
# Adders/Subtractors                                          : 31
 16-bit adder                                                 : 13
 4-bit adder carry out                                        : 4
 5-bit adder                                                  : 10
 5-bit adder carry out                                        : 3
 6-bit adder carry out                                        : 1
# Counters                                                    : 1
 4-bit up counter                                             : 1
# Registers                                                   : 113
 1-bit register                                               : 91
 16-bit register                                              : 8
 5-bit register                                               : 14
# Multiplexers                                                : 10
 1-bit 5-to-1 multiplexer                                     : 10
# Logic shifters                                              : 12
 16-bit shifter logical left                                  : 12

============================================================================
```

**Fig. 2.19** HDL synthesis report for 1-D 8-point DCT

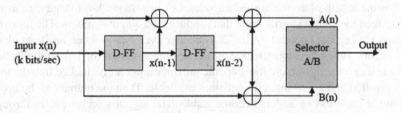

**Fig. 2.20** Block diagram of convolutional encoder for a rate ½., constraint length K = 3

Convolutional coding and Block coding are the two major forms of channel coding. Convolutional codes operate on serial data, one or a few bits at a time. Block codes operate on relatively large message blocks. There are a variety of useful convolutional and block codes, and a variety of algorithms for decoding the received coded information sequences to recover the original data. Convolutional encoding with Viterbi decoding is a FEC technique that is particularly suited to a channel in which the transmitted signal is corrupted mainly by Additive White Gaussian Noise (AWGN) [12].

The technique of convolutional coding transforms a binary message into a sequence of symbols to be transmitted. Upon reception, the received information must be related back to the original message bits. If there are no errors the process of decoding is readily accomplished. In general, convolutional coding techniques are applied to very long messages, such as the continuous stream of data from a satellite television transmitter.

A convolutional encoder with two shift registers is shown in Fig. 2.20.

**Table 2.3** State transition table for the convolutional encoder

| Current state | Output symbols, if input = 0 | Output symbols, if input = 1 |
|---|---|---|
| 00 | 00 | 11 |
| 01 | 11 | 00 |
| 10 | 10 | 01 |
| 11 | 01 | 10 |

The system block diagram can be expressed with the following equations:

$$A(n) = x(n) + x(n-1) + x(n-2)$$
$$B(n) = x(n) + x(n-2)$$

The basic building components of the convolutional encoder are flip-flops comprising the shift registers and Exclusive-OR gates comprising the associated Modulo-Two adders. The number of shift registers in the encoder generating the encoded sequence determines the capability of the decoder to detect and correct number of bit errors received on the receiver in the obtained encoded sequence of data.

In this encoder, data bits are provided at a rate of 'k' bits per second. Channel symbols are output at the rate of $n = 2k$ symbols per second. The constraint length $K = 3$ is the length of convolutional encoder, i.e., how many k-bit stages are available to feed the combinatorial logic that produces the output symbols. The input bit is stable during the encoder cycle. The encoder cycle starts when an input clock edge occurs. When the input clock edge occurs, the output of the left-hand flip-flop is clocked into the right-hand flip-flop, the previous input bit is clocked into the left-hand flip-flop and a new input bit becomes available. Then the outputs of the upper and lower modulo-two adders become stable. The output selector cycles through two states. In the first state, it selects and outputs the output of the upper modulo-two adder. In the second state, it selects and outputs the output of the lower modulo-two adder.

The state transition table that lists the channel output symbols, given the current state and the input data is shown in Table 2.3.

### 2.4.2 Viterbi Decoder

A Viterbi decoder uses the Viterbi algorithm for decoding bit stream that has been encoded using Convolutional codes. There are other algorithms for decoding a convolutional encoded stream (Ex: Fanon algorithm). The Viterbi algorithm is the most resource-consuming but it does the maximum likelihood decoding [12]. Viterbi decoding has the advantage that it has a fixed decoding time. It is well suited for hardware decoder implementation. But its computational requirements grow exponentially as a function of constraint length. So it is usually limited in practice to constraint lengths of $K \leq 10$.

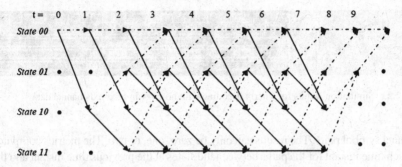

**Fig. 2.21** Trellis diagram for Viterbi decoding with encoder rate ½ and K=3

**Fig. 2.22** State transitions
from one state to the next
state

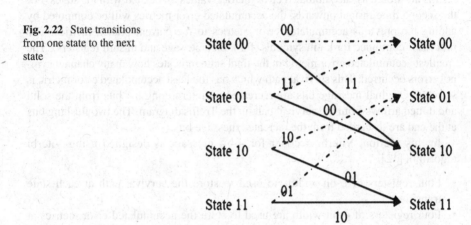

The most important concept to aid in understanding the Viterbi algorithm is the Trellis diagram. The Trellis diagram for the convolutional encoder rate ½, constraint length K=3 is shown in Fig. 2.21.

The four possible states of the encoder are depicted as four rows of horizontal dots. There is one column of four dots for the initial state of the encoder and one for each time instant during the message. For a 4-bit message with two encoder memory flushing bits, there are six time instants in addition to t=0, which represents the initial condition of the encoder. The solid lines connecting dots in the diagram represent state transitions when the input bit is a one. The dotted lines represent state transitions when the input bit is a zero. The expanded version of the transition between one time instant to the next is shown in Fig. 2.22. Notice the correspondence between the arrows in the Trellis diagram and the state transition diagram. Since the initial condition of the encoder is State 00, and the two memory flushing bits are zeros, the arrows start out at State 00 and end up at the same state [12].

Each time when a pair of channel symbols is received, the metric- Hamming distance between the received channel symbol pair and the possible channel symbol pairs is calculated for each state. The Hamming distance is computed by simply counting how many bits are different between the received channel symbol pair and the possible

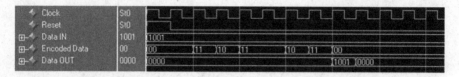

**Fig. 2.23** Simulation results for Viterbi decoding with no error in received channel data

channel symbol pairs. The results can only be zero, one, or two. The metrics computed at each time instant for the paths between the states at the previous time instant and the sates at the current time instant are called *branch* metrics. For the first time instant, the results are stored as "accumulated error metric" values associated with the states. For the second time instant onwards, the accumulated error metrics will be computed by adding the previous accumulated error metrics to the current branch metrics. The process is continued for k + m symbols (for k bits message and m shift registers). The smallest accumulated error metric in the final state indicates how many channel symbol errors occurred. This survival path which has the least accumulated error metric is selected. Original message bits are recreated by interpreting the bits from the solid and dotted arrows from the survival path in the Trellis diagram. The two flushing bits at the end are discarded from the recreated message bits.

In this section, Viterbi decoder for 4-bit message is designed using Viterbi algorithm [12].

- Four registers of 6-bit width are used to store the survival path at each state transition.
- Four registers of 4-bit width are used to store the accumulated error metrics at each state.
- At the end of the last state, the survival path having the least accumulated error metrics is used to reproduce the estimated input message bits from the survival path register.

### 2.4.3   Simulation and Synthesis Results

In this section, Convolutional encoder is designed using two shift-registers and Viterbi decoder is designed using Accumulated Error Metrics algorithm. The design is simulated using ModelSim®.

Assuming the input data to the convolutional encoder is x = (1001), the encoded sequence is, e = (11 10 11 11 10 11). Following different cases are simulated to test the Viterbi decoder design:

1. No error in the received data from the channel. The simulation result for this case is shown in Fig. 2.23.
   *Received data*: 11 10 11 11 10 11
2. One bit error in the received data from the channel. The simulation result for this case is shown in Fig. 2.24.

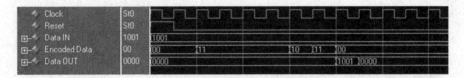

**Fig. 2.24** Simulation results for Viterbi decoding with one bit error in received channel data

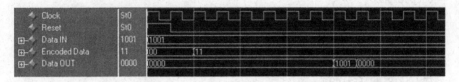

**Figure 2.25** Simulation results for Viterbi decoding with two bits error in received channel data

```
==========================================================================
HDL Synthesis Report

Macro Statistics
# ROMs                                              : 4
 4x10-bit ROM                                       : 2
 4x12-bit ROM                                       : 2
# Adders/Subtractors                                : 16
 3-bit adder                                        : 16
# Counters                                          : 1
 3-bit up counter                                   : 1
# Registers                                         : 17
 1-bit register                                     : 8
 3-bit register                                     : 4
 4-bit register                                     : 1
 6-bit register                                     : 4
# Comparators                                       : 13
 3-bit comparator greater                           : 4
 3-bit comparator lessequal                         : 9
# Multiplexers                                      : 22
 1-bit 4-to-1 multiplexer                           : 18
 3-bit 4-to-1 multiplexer                           : 4
# Xors                                              : 2
 1-bit xor2                                         : 2

==========================================================================
```

**Fig. 2.26** HDL synthesis report for convolutional encoder and Viterbi decoder

*Received data*: 11 11 11 11 10 11

3. Two bits error in the received data from the channel. The simulation result for this case is shown in Fig. 2.25.
   *Received data*: 11 11 11 11 11 11

This design is implemented on Xilinx® Spartan 2E FPGA. The HDL synthesis report is shown in Fig. 2.26.

# Appendix

## A. Verilog HDL code for PN sequence generator

```verilog
module pn_sequence(clk,rst,pnOUT);

input clk;
input rst;
output pnOUT;

// Instantiate PN sequence clock : 100 KHz
pn_clock pn_clock_seq(clk,rst,pnCLK);

// Generate PN Sequence : m = 4
reg pnOUT;
reg [3:0]shift4reg;

always @(posedge pnCLK, posedge rst)
begin
    if(rst)
    begin
        pnOUT <= 1'b0;
        shift4reg[3:0] <= 4'b0;
    end
    else
    begin
        shift4reg[0] <= ~(shift4reg[0] ^ shift4reg[3]);
        shift4reg[1] <= shift4reg[0];
        shift4reg[2] <= shift4reg[1];
        shift4reg[3] <= shift4reg[2];
        pnOUT <= shift4reg[3];
    end
end
endmodule

// PN Sequence Clock  : 100 KHz
module pn_clock(clk,rst,pnCLK);
input clk;
input rst;
output pnCLK;

reg pnCLK;
integer pnCLKCnt;

always @(posedge clk, posedge rst)
begin

    if(rst)
    begin
        pnCLK <= 1'b0;
        pnCLKCnt = 0;
    end
    else
    begin
        pnCLKCnt = pnCLKCnt + 1;
        if(pnCLKCnt >= 250)
        begin
            pnCLK <= !pnCLK;
            pnCLKCnt = 0;
        end
```

```
            else
                pnCLK <= pnCLK;
        end
    end
endmodule
```

## B. Verilog HDL code for LP FIR filter using Distributed Arithmetic Architecture

```
    module fir_filter(clk,rst,y,x0);

    input clk;
    input rst;
    input [9:0]x0;
    output [21:0]y;
    reg [21:0]y;
    reg [9:0]r1,r2,r3,r4,r5,r6,r7,r8,r9,r10,r11,r12,r13,r14,r15,r16;

    reg [12:0]temp;
    reg [21:0]calc;
    reg [3:0]count;
    reg load;


    always @(posedge clk, posedge rst)
    begin
        if(rst)
        begin
         r1<=0; r2<=0; r3<=0; r4<=0; r5<=0; r6<=0; r7<=0; r8<=0;
         r9<=0; r10<=0; r11<=0; r12<=0; r13<=0; r14<=0; r15<=0; r16<=0;
         temp = 0;
         calc = 0;
         count = 0; load = 1;
         y = 0;
        end
        else
        begin
            if(load)
            begin
                r1 <= x0;r2 <= r1;r3 <= r2;r4 <= r3;
                r5 <= r4;r6 <= r5;r7 <= r6;r8 <= r7;
                r9 <= r8;r10<= r9;r11<= r10;r12<= r11;
                r13<= r12;r14<= r13;r15<= r14;r16<= r15;
                load = 0; count = 0;
            end

        else
            begin
temp = block1Value( r1[count], r2[count], r3[count], r4[count]) +
        block2Value( r5[count], r6[count], r7[count], r8[count]) +
        block2Value(r12[count],r11[count],r10[count], r9[count]) +
        block1Value(r16[count],r15[count],r14[count],r13[count]);
```

```
        if((count == 4'b1001) &&
            (r1[count]  ||   r2[count]  ||   r3[count]  ||   r4[count]  ||
             r5[count]  ||   r6[count]  ||   r7[count]  ||   r8[count]  ||
             r12[count] ||   r11[count] ||   r10[count] ||   r9[count]  ||
             r16[count] ||   r15[count] ||   r14[count] ||   r13[count]))

            // For negative numbers
            calc = calc + (~(temp << count) + 1);
        else
            calc = calc + (temp << count);
            if(calc[12] == 1'b1)
                calc[21:13] = 9'b111111111;
            else
                calc[21:13] = 9'b000000000;

                if(count == 4'b1001)
                begin
                    y = calc;
                    calc = 0;
                    temp = 0;
                    load = 1;
                end
                else
                 count = count + 1;
        end
    end
end


function [12:0]block1Value; // LUT_1
input a1,a2,a3,a4;
begin
    case({a1,a2,a3,a4})
        4'b0000 : block1Value = 13'b0000000000000;
        4'b0001 : block1Value = 13'b1111111110100;
        4'b0010 : block1Value = 13'b1111111111110;
        4'b0011 : block1Value = 13'b1111111110010;
        4'b0100 : block1Value = 13'b0000000010110;
        4'b0101 : block1Value = 13'b0000000001001;
        4'b0110 : block1Value = 13'b0000000010011;
        4'b0111 : block1Value = 13'b0000000000111;
        4'b1000 : block1Value = 13'b0000000001000;
        4'b1001 : block1Value = 13'b1111111111100;
        4'b1010 : block1Value = 13'b0000000000111;
        4'b1011 : block1Value = 13'b1111111111011;
        4'b1100 : block1Value = 13'b0000000011100;
        4'b1101 : block1Value = 13'b0000000010001;
        4'b1110 : block1Value = 13'b0000000011010;
        4'b1111 : block1Value = 13'b0000000010000;
        default : block1Value = 13'b0000000000000;
    endcase

    //$display("blockValue 1 : %b\n",block1Value);
end
endfunction
```

```
function [12:0]block2Value; // LUT_2
input b1,b2,b3,b4;
begin
    case({b1,b2,b3,b4})
        4'b0000 : block2Value = 13'b0000000000000;
        4'b0001 : block2Value = 13'b0000010010100;
        4'b0010 : block2Value = 13'b1111111110011;
        4'b0011 : block2Value = 13'b0000010111000;
        4'b0100 : block2Value = 13'b1111111101110;
        4'b0101 : block2Value = 13'b0000010000010;
        4'b0110 : block2Value = 13'b1111110000000;
        4'b0111 : block2Value = 13'b0000001110100;
        4'b1000 : block2Value = 13'b0000000010110;
        4'b1001 : block2Value = 13'b0000011001001;
        4'b1010 : block2Value = 13'b0000000001000;
        4'b1011 : block2Value = 13'b0000010011000;
        4'b1100 : block2Value = 13'b0000000000100;
        4'b1101 : block2Value = 13'b0000010010110;
        4'b1110 : block2Value = 13'b1111111110100;
        4'b1111 : block2Value = 13'b0000010001010;
        default : block2Value = 13'b0000000000000;
    endcase
        //$display("blockValue 2 : %b\n",block2Value);
end
endfunction
endmodule
```

## C. Verilog HDL code for Convolutional encoder

```
module Conv_Encoder(clk, rst, dataIn, dataOut);

input clk;
input rst;
input [3:0]dataIn;
output [1:0]dataOut;

reg [1:0]dataOut;
reg [3:0]inBit;
reg ff1,ff2;


always@(posedge clk, posedge rst)
begin
    if(rst)
    begin
        dataOut <= 2'b0;
        ff1 <= 1'b0;
        ff2 <= 1'b0;
        inBit <= dataIn;
    end
    else
    begin
        ff1 <= inBit[0];
        ff2 <= ff1;
        dataOut[0] <= (inBit[0]^ff2);
        dataOut[1] <= (inBit[0]^ff1)^ff2;
        inBit <= inBit >> 1;
    end
end
endmodule
```

# References

1. Meel J (1999) Introduction to spread spectrum, Cirius Communications, Belgium
2. Miller A, Gulotta M (2004) PN generators (XAPP211), Xilinx Inc
3. An Introduction to Direct Sequence – Spread Spectrum (2003), Maxim Integrated Products Inc
4. Proakis JG, Manolakis DK (1995) Digital signal processing: principles, algorithm and application, 3rd edn. Prentice Hall, Englewood Cliffs
5. Booth's Algorithm: Multiplication and Division (2010) http://www.scribd.com/doc/3132888/Booths-Algorithm-Multiplication-Division. Accessed Oct 2010
6. Palinitkar S (2003) Verilog HDL: a guide to digital design and synthesis, 2nd edn. Prentice Hall, Palo Alto
7. Grover RS, Shang W, Li Q (2002) A faster distributed arithmetic architecture for FPGAs. In: ACM/SIGDA 10th International symposium on field-programmable gate arrays, Monterey, CA, USA, 24–26 Feb 2002, pp 31–39
8. Marshall D (2001) The discrete cosine transform. Cardiff Schoo of Computer Science & Informatics. http://www.cs.cf.ac.uk/Dave/Multimedia/node231.html. Accessed 10 October 2006
9. Khawan S, Baloch S, Pai A, Ahmed I, Aydin N, Arslan T, Westall F (2004) Efficient implementation of mobile video computations on domain-specific reconfigurable arrays. In: Conference on design, automation and test in Europe, vol 2, Paris, 16–20 Feb 2004, p 21230
10. Meyer-Baese U (2006) Digital signal processing with field programmable gate arrays, 2nd edn. Springer, Berlin/New York
11. Andraka Consulting Group, Inc. (2007) The CORDIC algorithm. http://www.andraka.com/cordic.htm. Accessed 2 April 2007
12. Fleming C (2006) A tutorial on convolutional coding with Viterbi decoding. Spectrum applications. http://home.netcom.com/%7Echip.f/viterbi/tutorial.html. Accessed 10 April 2006
13. Vahid F, Lysecky R (2007) Verilog for digital design. Wiley, Hoboken

# Chapter 3
# ASIC Design

The evolution in the VLSI industry contributing to the rapid technology changes, tremendous competition among vendors and demand in the market for ICs all these factors have led to consider the time to market factor with utmost importance. With maximum performance and least turnaround time, ASIC seems to be the best option to meet the ever growing demands for quality chips.

In this chapter, a comprehensive study on the ASIC design flow with various constraints is done along with an implementation of two simple systems to demonstrate the concept. SRAM architecture is designed and implemented using ASIC synthesis tools. Also, a Systolic Array Matrix multiplier is designed and modeled using Verilog HDL, Synthesized using Synopsys Design Compiler, Static Timing Analysis of the designs using Prime Time, Formal Verification using Formality and functional simulation of the synthesized net-list using ModelSim.

This chapter also demonstrates the Physical design process for Systolic Array Matrix multiplier. Synopsys Astro is used for the Physical design process. ModelSim and Prime Power are used as supplementary tools for power analysis of the design.

The pre-requisite to approach this chapter would be an adequate knowledge of ASIC design flow, concepts of physical design, CAD tools, Verilog HDL and basics of digital electronics.

## 3.1 ASIC Front-End Memory Design

### 3.1.1 Introduction

The explosive growth of the internet has increased the demand for high speed data communications systems that require fast processors and high-speed interfaces to peripheral components. While the processors in these systems have improved in performance, Static RAM (SRAM) performance has not kept pace. New SRAM architectures are evolving to support the throughput requirements of current systems [1]. Some of the well-known architectures are discussed in the following sections.

V.A. Chandrasetty, *VLSI Design: A Practical Guide for FPGA and ASIC Implementations*, SpringerBriefs in Electrical and Computer Engineering, DOI 10.1007/978-1-4614-1120-8_3, © Springer Science+Business Media, LLC 2011

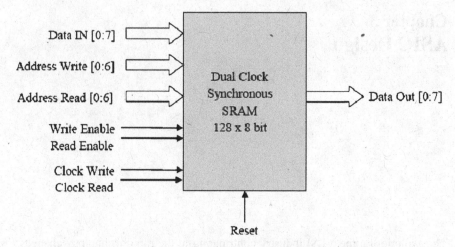

Data IN [0:7]

Address Write [0:6]

Address Read [0:6]

Write Enable
Read Enable

Clock Write
Clock Read

Dual Clock
Synchronous
SRAM
128 x 8 bit

Data Out [0:7]

Reset

**Fig. 3.1** Functional block diagram of dual clock synchronous SRAM architecture

### 3.1.2 Memory Architecture and Specifications

The Dual clock synchronous SRAM architecture uses two independent clocks with two different address buses for write and read operations [2]. The functional block diagram of this architecture is shown in Fig. 3.1.

The Dual clock synchronous SRAM architecture is used to increase the throughput of the system. Since two independent address buses is used for write and read operations controlled by two clocks, the read and write operations can be performed simultaneously and independently, hence enhancing the overall efficiency of the system in memory operations [3].

In this section, Dual clock synchronous SRAM architecture is chosen to design and implement the design. A memory bank of 128 bytes is designed with two synchronous read and write clocks, synchronous read and write enable control signal and a synchronous reset for the entire design. Data bus of 8 bits wide for input and output is used for write and read operations respectively. Address bus of 7 bits wide is used each for read and write operations.

### 3.1.3 Implementation and Simulations

The Dual clock synchronous SRAM is designed and implemented using Verilog HDL. Functional simulations are carried out using ModelSim. The design is synthesized using Synopsys Design Compiler [4] and Static Timing Analysis (STA) is done using Prime Time. Synopsis Design Constraints (SDC) file is also generated from Prime Time for this design. Formal verification on the generated net-list is performed using Formality tool.

The functional simulation of the design using ModelSim is shown in Fig. 3.2.

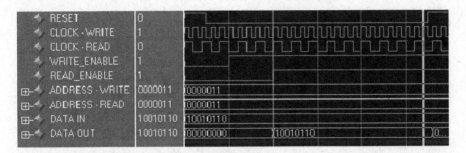

**Fig. 3.2**  Simulation of dual clock synchronous SRAM

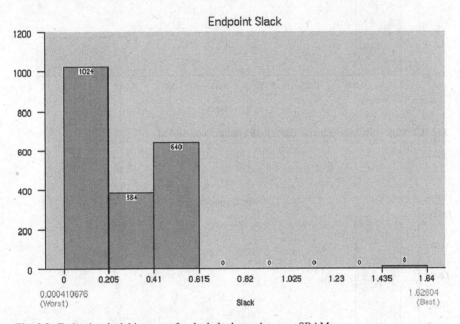

**Fig. 3.3**  End point slack histogram for dual clock synchronous SRAM

### 3.1.4   Results Analysis and Conclusion

The dual clock synchronous SRAM design is loaded to Prime Time for Static Timing Analysis. Following results have been analyzed for the design:

1. The end point slack analysis for the design with the selection of a maximum of 100 endpoints and 8 bins is represented by the histogram shown in Fig. 3.3.
2. The path slack analysis for the design with the selection of a maximum of 100 paths and 8 bins is represented by the histogram shown in Fig. 3.4.
3. The net capacitance analysis for the design with the selection of a maximum of 100 nets and 8 bins is represented by the histogram shown in Fig. 3.5.

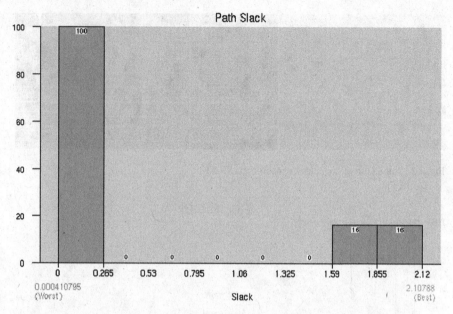

**Fig. 3.4**  Path slack histogram for dual clock synchronous SRAM

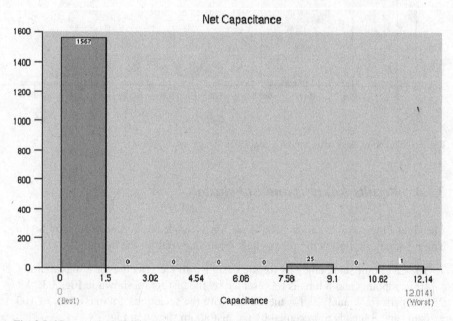

**Fig. 3.5**  Net capacitance slack histogram for dual clock synchronous SRAM

The reports obtained from the synthesis and static timing analysis of the design are as follows:

1. *Timing Analysis*
   Clock Read = 3.2 ns with Setup Slack = 0 ns and Hold Slack = 0.86 ns
   Clock Write = 2 ns with Setup Slack = 0 ns and Hold Slack = 0.99 ns
2. *Area Report*
   Total Area = 131324.37 μm²
   Combinational Area = 121873 μm²
   Sequential Area = 7575.50 μm²
   Net Interconnect Area = 1875.87 μm²
3. *Power Report*
   Total Dynamic Power = 917.97 mW
   Cell Internal Power = 328.32 mW
   Net Switching Power = 589.64 mW
   Cell Leakage Power = 68.81 μW
4. *Components Report*
   Number of Ports = 35
   Number of Nets = 1593
   Number of Cells = 1559
   Number of References = 39

*Conclusion*:

The Dual clock synchronous SRAM is designed for a memory bank of 128 bytes only. The same design can be enhanced for larger memory bank using the same architecture. The existing architecture makes use of two independent read and write clocks which increases the throughput compared to the traditional single clock architectures. This Dual clock architecture can be further enhanced by using a single clock with positive edge triggering for read operation and negative edge triggering for write operation, making the design to work with a single clock. But in this case, extra clock period is provided for write operation which may not be necessary.

## 3.2   ASIC Front-End Matrix Multiplier Design

### 3.2.1   Introduction

The computational speed greatly matters in high-end designs where multiplication is incorporated. As multiplication is one of the high resource consuming process, the matrix multiplier is one such process which involves multiplication. Various architectures and designs are proposed in order to optimize the efficiency of the multipliers. This section discusses on design and implementation of one such matrix multiplier architecture.

### 3.2.2   Problem Statement

In this section a matrix multiplier is designed and implemented with the following specifications:

- Systolic Array architecture is used to design the multiplier
- Single clock is used to control the entire design
- Two matrices of order $3 \times 3$ to be multiplied is designed
- The matrix multiplier design is hierarchical
- The data width is four for each of the input matrix elements
- The multiplier accepts the data stored in memory
- A positive slack of around 15% of the clock is ensured
- Full Scan Chain DFT methodology is incorporated to make the design Observable and Controllable

### 3.2.3   Matrix Multiplier Design

The Systolic Array architecture is used to design the $3 \times 3$ matrix multiplier system. This architecture consists of Data Processing Units (DPU) arranged in the form of an array. The DPU is nothing but a Multiplier and Accumulate (MAC) unit which processes each data entering the system. This kind of architecture incorporates parallel processing and pipelining mechanism, hence increasing the throughput and latency of the system [5]. The functional block diagram of Systolic Array matrix multiplier is shown in Fig. 3.6.

The schematic of Systolic array blocks generated by Synopsys Design Compiler is shown in Fig. 3.7. The matrix A and B that needs to be multiplied is fed in to the multiplier with row and columns of the matrices arranged with single clock delays. At the end of 7 clock cycles, the value in the accumulator of DPUs itself is the final multiplied values of matrix A and B. Hence the latency of the system is 7 clock cycles.

### 3.2.4   Implementation and Simulations

The Systolic array matrix multiplier of order $3 \times 3$ is designed and modeled using Verilog HDL. The pre-synthesis functional verification of the design is simulated and tested using ModelSim. The synthesis is carried out using Synopsys Design Compiler (DC) [6]. A script is used to automate DC for synthesis process. Full multiplexed scan DFT is incorporated to make the system Testable, Controllable and Observable. The synthesized design is ported to Prime Time for Static Timing Analysis. For the optimized design obtained, Synopsys Design

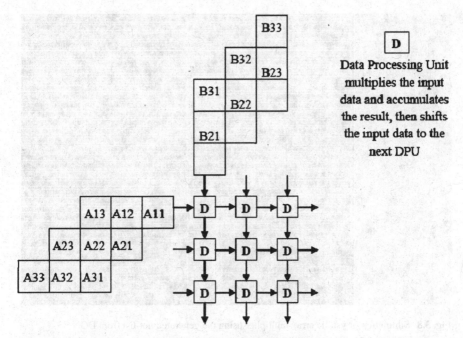

Fig. 3.6 Functional block diagram of systolic array matrix multiplier

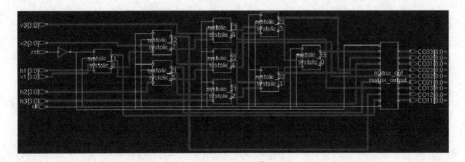

Fig. 3.7 Schematic of systolic array blocks generated by synopsys design compiler

Constraints (SDC) file and Verilog net-list is generated using DC. Formal verification of the generated net-list across the designed Verilog code is done using Formality tool. The verified net-list is then finally simulated for functional verification. The functional simulation of the generated net-list for the design using ModelSim is shown in Fig. 3.8.

From the Fig. 3.8 it can be noted that the inputs to the system are $3 \times 3$ matrices A and B, clock, reset and DFT inputs test_si and test_se. The functional verification of the net-list is carried out for the following input vectors and the simulated output

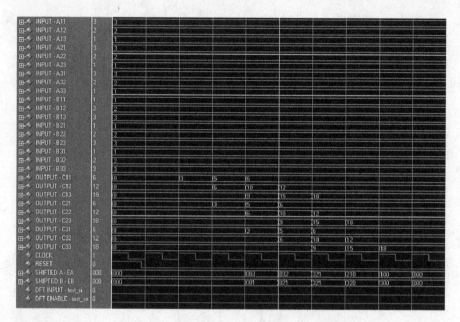

**Fig. 3.8** Simulation of ystolic array multiplier using the generated net-list from DC

C is verified. The DFT scan is disabled by forcing the input of DFT test input and enable signal to zero. The output ports EA and EB are used to register elements of matrices A and B that are pushed out of the systolic array after the multiplication. These ports may be extended to increase the order of the array and also used for debugging the system.

$$A \times B = C$$

$$\begin{bmatrix} 3 & 2 & 1 \\ 3 & 2 & 1 \\ 3 & 2 & 1 \end{bmatrix} \times \begin{bmatrix} 1 & 2 & 3 \\ 1 & 2 & 3 \\ 1 & 2 & 3 \end{bmatrix} = \begin{bmatrix} 6 & 12 & 18 \\ 6 & 12 & 18 \\ 6 & 12 & 18 \end{bmatrix}$$

### 3.2.5  Analysis of Results and Conclusion

The Systolic Array matrix multiplier design is loaded to Prime Time for Static Timing Analysis. Following results have been analyzed for the design:

1. The end point slack analysis for the design with the selection of a maximum of 100 endpoints and 8 bins is represented by the histogram is shown in Fig. 3.9.
2. The path slack analysis for the design with the selection of a maximum of 100 paths and 8 bins is represented by the histogram is shown in Fig. 3.10.

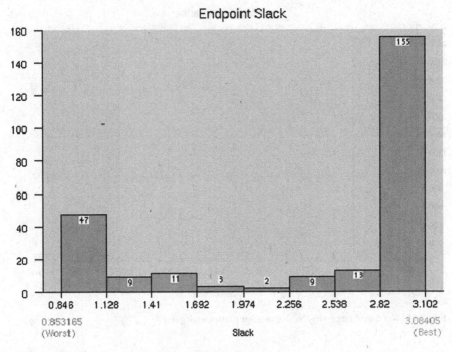

**Fig. 3.9** End point slack histogram for matrix multiplier

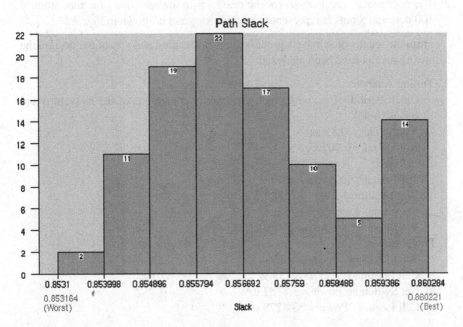

**Fig. 3.10** Path slack histogram for matrix multiplier

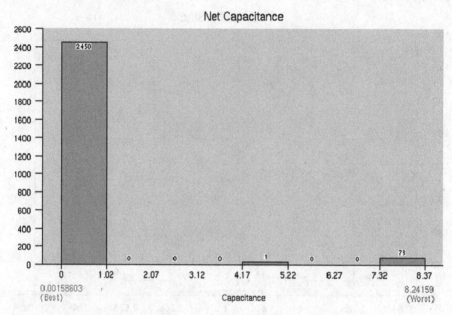

**Fig. 3.11** Net capacitance slack histogram for matrix multiplier

3. The net capacitance analysis for the design with the selection of a maximum of 100 nets and 8 bins is represented by the histogram is shown in Fig. 3.11.

From the results of synthesis process and static timing analysis of the design, the following results have been analyzed:

1. Timing Analysis
   Clock Period = 4 ns, with clock uncertainty constraints of 0.3 ns (setup) and 0.2 ns (hold)
   Setup Slack = 0.85 ns
   Hold Slack = 0.30 ns
2. Area Report
   Total Area = 620566.37 μm²
   Combinational Area = 616889.25 μm²
   Sequential Area = 2084.75 μm²
   Net Interconnect Area = 1592.35 μm²
3. Power Report
   Total Dynamic Power = 1.2764 W
   Cell Internal Power = 1.2733 W
   Net Switching Power = 3.1787 mW
   Cell Leakage Power = 53.4279 μW

4. Test Coverage Report
   Test Coverage = 100% for Full Scan Multiplexed DFT
   Total Faults = 15078
   Detectable Faults = 14997
   Undetectable Faults = 81
5. Components Report
   Number of Ports = 182
   Number of Nets = 2524
   Number of Cells = 2422
   Number of References = 90

*Conclusion*:

The matrix multiplier design can be optimized for better power, area and timing performances by incorporating DFT in RTL design itself instead of using DFT flip-flops. The design can also be extended from $3 \times 3$ to a higher order by reusing the systolic data processing unit in the chain of arrays. For the current design, it is also ensured to have a setup slack of 0.85 ns to take care of uncertainties in the physical design and fabrication process.

## 3.3 Physical Design of Matrix Multiplier

### 3.3.1 Introduction to Systolic Array Matrix Multiplier

The Physical design of the Systolic array matrix multiplier design is carried out in this section. Various inputs and configurations are required in the physical design flow to obtain error free and optimized layout of the design.

The physical design process requires information of:

- *Standard cells*: A standard cell is a group of transistor and interconnects structures, which provides a Boolean logic function such as, NAND, NOR, Inverters, etc. or a storage function like flip-flop or latch
- *IO cells*: The IO cell consists of Input and Output circuits (pads) to interface with the core logic and external world
- *Special cells*: These cells are macros to serve special purpose such as memory, PLL, etc.

All these library cells are technology dependent. The technology file is an important input to the physical design process. It consists of following parameters:

- Metal Layer definitions
- Via definitions
- Process design rules (minimum width, spacing, etc.)

**Table 3.1** Metal layer characterization in 130 nm technology

| Metal layer | Metal layer ID | Alignment | Color |
|---|---|---|---|
| 1 | 14 | Horizontal | Blue |
| 2 | 18 | Vertical | Yellow |
| 3 | 22 | Horizontal | Red |
| 4 | 26 | Vertical | Green |

- TLU parasitic capacitance models
- Preferred routing directions
- GUI display info (color and fill of layers)
- Units (time, capacitance, distance)

In Synopsys Astro, technology file is recognized in ".tf" format and in Cadence SOC Encounter it is in ".lef" format. In this section, Synopsys Astro with 130 nm technology is used to implement the matrix multiplier design.

Depending on the technology, the number of metal layers is also defined. Metal layer characterization is absolutely necessary to perform clean routing of cells with least congestion [7]. The characterization for metal layers in 130 nm technology is shown in Table 3.1.

The Physical design process requires certain basic inputs in-order to generate desired results. The list of inputs required is as follows:

1. Verilog netlist for the design (*.v)
2. Timing libraries (*.lib)
3. Technology file (*.tf or *.lef)
4. TDF / IO assignment file (*.io or *.tdf)
5. Timing constraints file (*.sdc)

The Physical design process generates certain outputs and reports to analyze the design. Some of the most important outputs/reports generated are as follows:

1. Post layout Verilog netlist
2. SDF
3. SPEF
4. DSPF
5. SPICE
6. LEF/DEF
7. GDS II
8. Timing reports
9. Skew reports
10. DRC/ERC/LVS reports

The physical design process flow consists of various steps [8]. The detailed flow is shown in Fig. 3.12.

**Fig. 3.12** Physical design
flow

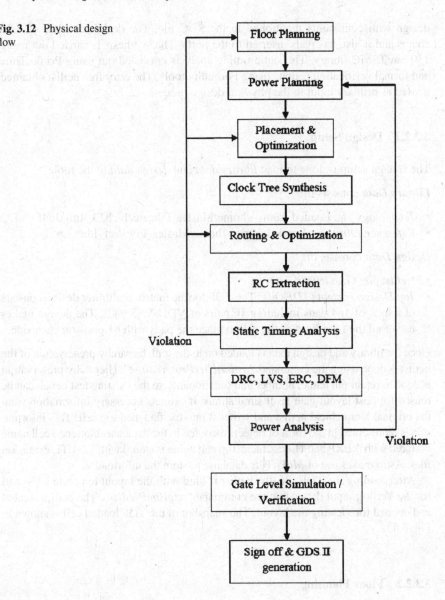

## 3.3.2   Physical Design Flow

The Physical design for the Systolic Array matrix multiplier is done using Synopsys
Astro [9]. The procedure is discussed in the following sections. A $3 \times 3$ 4-bit matrix
multiplier is designed with Systolic array architecture using Verilog and the optimized netlist is generated using Design Compiler. The netlist generated is a flatten

design with constraints mentioned in the SDC file. The design consists of scan chains and it also has pads inserted to the ports. The synthesis is carried out using 130 nm TSMC library. The static timing analysis is carried out using Prime Time and formal verification is done using Formality tools. The error free netlist obtained is used as primary input to the physical design process.

### 3.3.2.1   Design Setup

The Design setup is done to load *library data* and *design data* to the tool.

***Library Data*** consists of:

- *Technology file*: Loaded from – /home/Master_Files/tech/cb13_4m_tlu.tf
- *Reference Libraries*: Loaded from – /home/Master_Files/ref_lib/*.lib

***Design Data*** consists of:

- *Netlist file*: Gate level design
- *Top Design Format (TDF) file*: The TDF for the matrix multiplier design consists of a total of 164 pads including 16 pairs of VDD-VSS pads. The design makes use of all the four sides of the cell to place the pads with 41 ports on each side.

Once the library and design data is loaded on to the tool, hierarchy preservation of the netlist is done using the command "*astInitHierPreservation*". Hierarchy preservation is done to retain pin name, number and functionality so the existing test bench can be reused for post layout gate level simulations. It extracts necessary information from the original hierarchical netlist and writes it into the flattened top cell. This information is represented in the form of objects recorded in the database. Flattened cell name is stored with *.EXP and Hierarchical Top cell name is stored with *.NETL extension files. Astro makes use of *Milky Way* database to store the information.

After loading the TDF file, the netlist is bind with the layout to create a top cell for the Verilog input data using the command "*axgBindNetlist*". The cell is created and opened for viewing the layout. The snapshot of the TDF loaded cell is shown in Fig. 3.13.

### 3.3.2.2   Floor Planning

In the floor planning stage, following setups and configurations are done:

- Creating core and pad area
  - ○ Core Utilization = 0.6, which is the ratio of area of the core to total area of the cell. This is selected as per the assignment specification.
  - ○ Aspect ratio = 0.65, which is the ratio of Height to the Width of the core.
  - ○ Core Aspect = 1.
  - ○ Core to Pad distance is maintained with 60 μm.

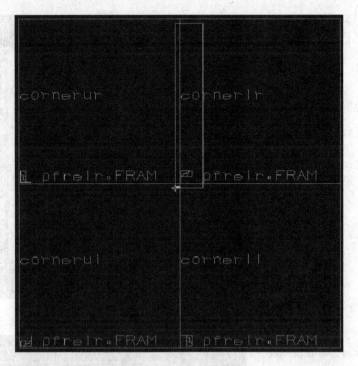

**Fig. 3.13** Snapshot of TDF loaded cell

- Creating standard cell rows
  - Row to Core ratio = 1, It is the ratio of total row area to the area of the core.
    A value of ≥ 1 is used for channel-less rows.
    A value of < 1 is used for rows with channel.
  - Horizontal row is selected to obtain rows aligned horizontally for placement of cells.
  - Flip first row double back option is used to utilize the power rails by merging them and even to save the area.

- Placing Macros in the core area
  The macro available in the design is placed in to the core area in this stage. In the matrix multiplier design, there are no macros. Hence this step is ignored.

The matrix multiplier design consists of elements, as reported by Astro at the floor planning stage:

- No. of signal ports: 132
- No. of Nets: 2237
- No. of Ports: 2203

**Fig. 3.14** Snapshot of cell after floor planning stage

**Fig. 3.15** Snapshot of alignment of rows in the core area

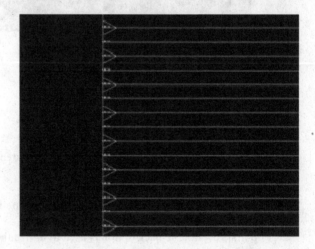

The snapshot of the cell after floor planning stage is shown in Fig. 3.14.

The snapshot of the core area of the cells with a close-up view of the alignment of the rows- flip first and double back is shown in Fig. 3.15.

### 3.3.2.3 Power Planning

Power planning is the stage where Power/Ground network is implemented in the design. The Systolic Array matrix multiplier is a flattened design without any macros.

Hence bottom-up approach is followed in power planning. Power planning consists of following steps:

- *Power Budget*
  The estimation of total dynamic core power is computed using VCD file and Prime Power. The method followed to do the same is shown in Fig. 6.6.
  Total dynamic power for the design $= 474.6$ mW
  Operating voltage for 130 nm technology $= 1.08$ V
  Total dynamic core current $= 440$ mA
- *Power/Ground Pads*
  The design consists of 132 ports. All the four sides of the cell are made use for port distribution. Considering that VDD/GND pairs of pad are required for every eight signal pads for a normal design, the number of VDD/GND pairs of pad required for Matrix multiplier design is,
  VDD/GND pairs of pad $=$ Total number of signal pads $(132)/8 \approx 16$
  Four pairs of VDD/VSS pads are incorporated on each side of the cell.
- *Pad to Core Trunk Width*
  The pad to core trunk width is given by,
  $W_{pc} =$ (Total dynamic core current) $\div$ (No. of sides $\times J_{max}$)
  Where, $J_{max} =$ Maximum current density of the metal
  $J_{max}$ for Metal layer $3 = 19.3$ mA/$\mu$m
  $J_{max}$ for Metal layer $4 = 49.5$ mA/$\mu$m
  $W_{pc}$ for Metal layer $3 = 5.7$ $\mu$m
  $W_{pc}$ for Metal layer $4 = 2.2$ $\mu$m
- *Pad and Core Ring*
  The pad ring is added to the design in-order to make the power/ground connections for IO pads. The core ring with power/ground connections are made with a metal width of 2 $\mu$m and offset from the core of 1 $\mu$m.
- *Straps/Trunks placement*
  In Systolic Array matrix multiplier design, since macros are not there, straps or trunks are incorporated. Only power/ground rails (axgPrerouteStandardCells) are used for connecting the standard cells.

The snapshot of cell after power planning stage is shown in Fig. 3.16.

### 3.3.2.4 Timing Setup

In the Timing setup stage, the timing information is provided to the tool to optimize placement and routing with required timing. It also performs Static Timing Analysis for the timing constraint applied to the design. The SDC is loaded to the tool using the command "*ataLoadSDC*".

Astro uses a congestion-based coupling model and TLU + capacitance tables to accurately model the pre-routing capacitance for nets prior to routing.

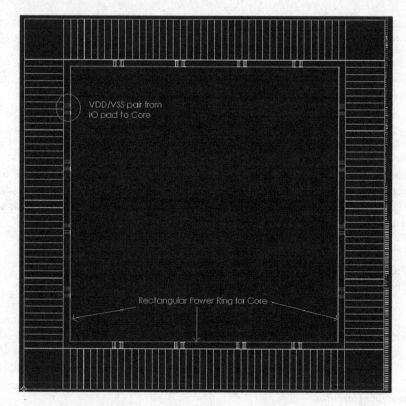

**Fig. 3.16** Snapshot of cell after power planning stage

This methodology eliminates the questionable derivation of the linear capacitance co-efficient and improves the accuracy of the model by taking into account increases in coupling capacitance due to increased congestion in different regions of the chip. The ITF is converted to TLU+format using the command "*cmItfToTLUPlus*". The snapshot of ITF to TLU+conversion is shown in Fig. 3.17.

Additional timing setup is carried out using the command "*atTimingSetup*". The following options are used in the timing setup for Matrix multiplier design:

1. Environment Setup
    Default options
2. Optimization Setup
    Target Setup Slack = *0.9*
    Target Hold Slack = *0.2*
3. Library Setup
    Default options
4. Parasitic Setup
    Parasitic Source: LPE

```
cmItfToTLUPlus
#t

--------- Sanity Check on TLU+ Inputs -------------
INFO: ResModel1 is detected
1. Checking the conducting layer names in ITF and mapping file ...
[ Passed! ]
2. Checking the via layer names in ITF and mapping file ...
[ Passed! ]
3. Checking the consistency of etch values between MW-tech and ITF ...
[ Passed! ]
4. Checking the consistency of Min Width and Min Spacing between MW-tech and ITF ...
[ Passed! ]
5. Checking the consistency of conducting layer thicknesses between MW-tech and ITF ...
[ Passed! ]
----------------- Check Ends -------------------
INFO: CapModel (/home/students/student/PEPS_FT_06_lab_work/Batch_2_Vikram/PD/Matrix_PD/
matrix/lib_2) is attached
INFO: ResModel (/home/students/student/PEPS_FT_06_lab_work/Batch_2_Vikram/PD/Matrix_PD/
matrix/lib_2) is attached
INFO: ITF_Combo (/home/students/student/PEPS_FT_06_lab_work/Batch_2_Vikram/PD/Matrix_PD
/matrix/lib_3) is replaced
ITF to TLU+ conversion successful !
```

**Fig. 3.17**  Snapshot of ITF to TLU+ successful conversion

```
ASTSUM: Summary of timing analysis (w/o xtalk)
ASTSUM:        Setup (Target=0.0000)        Hold          Num        Num
ASTSUM:       Slack    Num       Total     Slack    Num   MaxTrans MaxCap
ASTSUM:     -400.591    153    -18477.0    3.949      0     5485     1747
@@@ Total CPU      Time =     0:00:09
@@@ Total Elapsed Time =     0:00:09
@@@ Peak  Memory  Used =    93.55 MB

Timing Report OK
```

**Fig. 3.18**  Snapshot of timing report after loading the SDC for the design

     LPE mode: *Auto*

     Operating condition: *Max*

     Capacitance model: *TLU+*

5. Model Setup

     Operating condition: *Max*

     Net Delay Modes: *Medium Effort*

6. Xtalk Setup

     Default options, as noise or cross talk is not modeled in this design.

The snapshot of timing report after loading the raw SDC is shown in Fig. 3.18. The report shows that both setup and hold slack are violated. In further stages optimization needs to be done to meet the slack.

**Fig. 3.19** Placement process
flow after floor planning

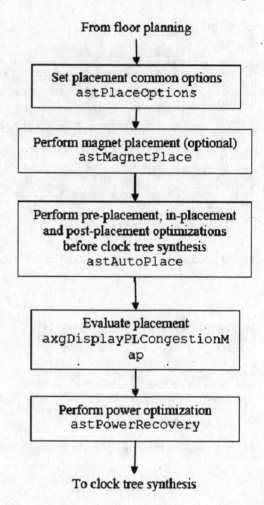

From floor planning

Set placement common options
`astPlaceOptions`

Perform magnet placement (optional)
`astMagnetPlace`

Perform pre-placement, in-placement
and post-placement optimizations
before clock tree synthesis
`astAutoPlace`

Evaluate placement
`axgDisplayPLCongestionM
ap`

Perform power optimization
`astPowerRecovery`

To clock tree synthesis

### 3.3.2.5   Placement

In the Placement stage, the standard cells are placed in the core area. The placement
process flow after floor planning is shown in Fig. 3.19 [8].

The common options for Placement are selected with the following modes or
constraints. The Astro commands used for each of the Placement options are provided in the brackets.

- *Optimization modes*:

    - *Congestion*: Distributes cell placement for minimum congestion,
    - *Timing*: Places cells to meet timing requirements.

```
ASTSUM: Summary of timing analysis (w/o xtalk)
ASTSUM:        Setup (Target=0.0000)        Hold        Num     Num
ASTSUM:        Slack    Num    Total    Slack    Num  MaxTrans MaxCap
ASTSUM:       -0.324    20     -3.0    0.069      0     197      32
@@@ Total CPU      Time =     0:00:08
@@@ Total Elapsed Time =     0:00:09
@@@ Peak  Memory  Used =   165.27 MB

Timing Report OK
```

**Fig. 3.20** Snapshot of timing report after placement and optimization stage

- *Location Constraints*:

  - *Consider pre-route types*: PG ring, PG pin; Astro recognizes the pre-routed nets of the types selected.
  - *No cells under pre-route of M3 & M4*; Astro doesn't place cells under the pre-routed nets on the metal layers selected.
  - *No cells under via V34*; Astro doesn't place cells under the vias specified.
  - *Short checking at pre-route of M3 & M4*; Astro doesn't place cells under the pre-routed nets on the layers specified if a short occurs or if it cannot access pins.

After the placement common options are selected, actual placement of cells is carried out in three different steps before CTS. The steps followed are discussed below:

- *Pre-Placement (astPrePS)*
  The Pre-Placement Optimization performs overall timing improvement. The goal of pre-place optimization is to correctly setup the design for placement of cells. This includes the handling of high fan-out nets, design cleanup and some optimization.
- *In-Placement (astAutoPlace)*
  The standard cell instances are actually placed in the core for the design. It follows the optimized placement solution obtained in Pre-placement optimization. Search and Refine option may be used to improve the cell placement by evaluating the current placement, determining congestion in the design and by changing the placement of cells within congestion areas.
- *Post-Placement (astPostPS1)*
  Post-Placement Optimization (PPO1) is performed after placement of cells. Additional optimization techniques are used to obtain best results. The goal of this step is to clean up some high fanout nets after placement, to fix timing constraints such as maximum capacitance, transition and also to prevent crosstalk.

  The snapshot of timing report after Placement and optimization stage is shown in Fig. 3.20. The report shows that hold slack is positive but setup slack still violates. An evident improvement in setup slack can be noticed as compared to that of SDC loaded report.

The snapshot of the cell after the Placement stage is shown in Fig. 3.21.

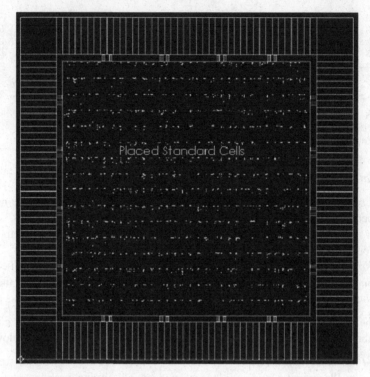

Placed Standard Cells

**Fig. 3.21** Snapshot of cell after placement stage

### 3.3.2.6   Clock Tree Synthesis

In this stage, the clock tree is synthesized to meet the timing requirements and avoid blockages and correlation problems between pre-routing and post-routing. Clock Tree Synthesis (CTS) is basically done for zero skew. It may add multi-level buffer trees according to the clock specification – skew and insertion delay. The CTS process may result in more buffers added, movement of cells, increase in congestion and even introduction of new timing and max capacitance/transition violations. The flow for CTS process is shown in Fig. 3.22 [8].

The CTS common options (astClockOptions) are selected before actually CTS is performed for the Matrix multiplier design. The options selected are as follows:

- Conditions: *worst*
- Skew type: *Global*
- Synthesis Effort: *Two*
- Gated clock tree: *True*
- Clock nets: "clk"; added from the loaded SDC (root clock)
- Target skew and insertion delay options are untouched as it takes from SDC

**Fig. 3.22** Clock tree synthesis flow after placement

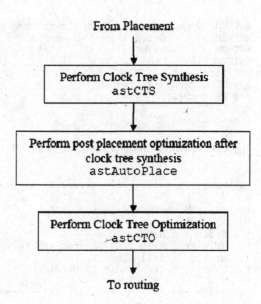

The clock tree is marked to set and propagate variable route rules before running CTS. This is done to prevent slacks from being disturbed during CTS. The following options are selected for marking clock tree (*astMarkClockTree*):

- Clock net name: *clk*
- Fix – clock tree and flip flops

Once the CTS common options are selected and the clock tree is marked, the design is ready for CTS (*astCTS*). The following options are selected to perform CTS:

- Conditions: *worst*
- Skew type: *Global*
- Design level: *Block*

After performing CTS, Post-Placement Optimization (PPO2) is done to obtain better timing. Post-Placement Optimization has several differences from Pre-Placement Optimization. During Post-Place Optimization, the placement engine is still active in order to accurately take into account any changes in the design. This is important when cells are either sized or moved, as this impacts the timing of the design based on the new placement information.

All of the optimization techniques employed during post-place optimization takes into account all sizing, cell-moving, cell-bypassing, buffer and inverter insertions, gate-duplication and net-splitting. Logical-remapping and area recovery are optional techniques which can be added to PPO. The following options are selected for PPO2 (*astPostPS*):

- *Setup Fixing*: to fix the setup slack violation
- *Max Tran Fixing*: to fix maximum transition violation

```
ASTSUM: Summary of timing analysis (w/o xtalk)
ASTSUM:       Setup  (Target=0.0000)        Hold        Num      Num
ASTSUM:       Slack     Num     Total    Slack    Num  MaxTrans MaxCap
ASTSUM:      -0.069      10     -0.5     0.071      0      0       24
@@@ Total CPU     Time  =    0:00:08
@@@ Total Elapsed Time  =    0:00:09
@@@ Peak  Memory  Used  =  166.66 MB

Timing Report OK
```

**Fig. 3.23** Snapshot of timing report after CTS and CTO

```
*********************************************************************
*
*    Clock Tree Skew Reports
*
*    Tool    : Astro
*    Version : X-2005.09-SP4 for AMD.64 -- Apr 19, 2006
*    Design  : cell_CTS
*    Date    : Thu Jun  7 13:06:21 2007
*
*********************************************************************



======== Clock Global Skew Report ==================================

Clock:  clk
Pin:    clk
Net:    clk

Phase Type               = rise && fall
Operating Condition      = worst
The clock global skew    = 0.086
The longest path delay   = 1.914
The shortest path delay  = 1.828
```

**Fig. 3.24** Clock global skew report for the matrix multiplier design

- *Max Cap Fixing*: to fix maximum capacitance violation
- *Logic-Remapping*: used during setup slack optimization. It also attempts to reduce the number of stages in a critical path for overall timing improvement.

Even after PPO2, the setup slack was violated. In order to eliminate/reduce the violation, "*pdsCROptimization*" command is used. It performs timing optimization to further reduce the total negative slack of the design to isolate the most critical paths which are hard to optimize. The Clock Tree Optimization (*astCTO*) is carried out to reduce or pull down the skew to zero.

The snapshot of timing report after CTS and CTO for Matrix multiplier design is shown in Fig. 3.23. From the report it can be inferred that only setup slack and max capacitance violation exists.

The clock global skew analysis report for the Matrix multiplier design is shown in Fig. 3.24.

**Fig. 3.25** Routing flow after CTS

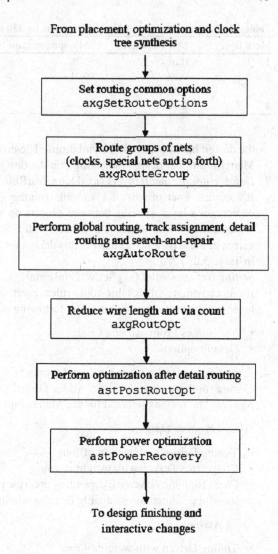

From placement, optimization and clock tree synthesis

↓

Set routing common options
axgSetRouteOptions

↓

Route groups of nets
(clocks, special nets and so forth)
axgRouteGroup

↓

Perform global routing, track assignment, detail routing and search-and-repair
axgAutoRoute

↓

Reduce wire length and via count
axgRoutOpt

↓

Perform optimization after detail routing
astPostRoutOpt

↓

Perform power optimization
astPowerRecovery

↓

To design finishing and interactive changes

### 3.3.2.7 Routing

In the Routing stage, the metal layers are drawn for all interconnects complying Design Rule Checks (DRC). It is also made sure that circuit timing, clock skew, signal net transition and capacitance limits are maintained in acceptable limits. The design flow used in the Routing process is shown in Fig. 3.25 [8].

The following procedure is followed for routing the Matrix multiplier design:

1. *Check design for Route* (axgCheckDesignForRoute)
   It performs a check for optimization in order to substantiate any errors in the design that might need to be fixed. It checks for pin access points, cell instance wire tracks, pin out of boundaries, min-grid and pin design rules and blockages to ensure they meet the design requirements. An error cell will be generated if

**Table 3.2** Typical values of metal layer parameters for 130 nm technology

| Metal layer | Min width (μm) | Min spacing (μm) | Min area (μm²) | Pitch (μm) |
|---|---|---|---|---|
| 1 | 0.16 | 0.18 | 0.122 | 0.41 |
| 2 | 0.20 | 0.21 | 0.144 | 0.41 |
| 3 | 0.20 | 0.21 | 0.144 | 0.515 |
| 4 | 0.44 | 0.46 | 0.562 | 0.97 |

the design has some errors or violations. The error cell wasn't generated for the Matrix multiplier design indicating that the design is error free.

2. *Define rules for metal layers* (axgDefineVarRule)

It specifies a set of rules for variable routing and defines minimum width of objects on a layer, spacing between objects on a layer and size of array to use with a contact. The values are verified with that in the technology file. Typical values of these parameters used in this design for 130 nm technologies are shown in Table 3.2.

3. *Setting Net constraint* (axgSetNetConstraint)

It sets constraints of variable-route rules, layer, timing-driven, spacing, and top-layer probe constraints for nets. The following options are set for the design:

- Net Names From: *All clock nets*
- Default options

4. *Setting Route Options* (axgSetRouteOptions)

Certain routing options are selected for Global routing, Track Assign and Detail routing. The options selected for the Matrix multiplier design is as follows:

*Global Routing Options*

- Timing Driven with weight of four
- Congestion Driven with weight of four
- Clock Routing: *Balanced*; depending upon the pin distribution of each clock net, the global router automatically decides whether or not to use single-trunk

*Track Assign*

- Timing Driven with weight of one

*Detail Routing*

- Connect open nets
- Timing Driven
- Single-row/column via array: *center*; places the contact center at the corner where the router changes routing layers. This result in a "T" shaped corner

*Library cells and Design Rules*

- Poly Pin Access: *auto*; connects poly pins if poly pins exist in the design
- DRC Distance: *Manhattan*; checks in the X and Y directions. Both X and Y spacing must be greater than the minimum spacing rule
- Same Net Notch: *check and fix*; Attempts to fix same net notch violations

- Fat Wire Checking: *merge then check*; Tries to merge thin wires to form fat wires for fat-wire spacing rule checking
- Merge Fat wire on: *signal routing too*; Merges all types of wires to form fat wires
- Wire/Contact End-of-line Rule: *check and fix*; instructs the router to not connect to "cross-vias" with wrong way wires to avoid end-of-line rule violations

5. *Routing Net Group* (axgRouteGroup)

It Routes the nets specified in this group. The routing replaces any previous routing. In this design clock net is selected for routing. The following options are used to route the clock nets:

- Net Names from: *All clock nets*
- Phase: global, track assign, detail
- Search Repair Loop: *Five*
- Dangling wires: *discard*; the router discards or removes all the dangling wires or contacts before starting to connect nets that are broken.
- Optimize routing pattern

The snapshot of the cell showing the routed clock net from the IO pad to the core logic is shown in Fig. 3.26.

6. *Automatic Routing* (axgAutoRoute)

Automatic routing is done to sequentially run Global routing, Track assignment, Detail routing and Search & Repair steps followed by Post-Route Optimization to optimize the detail routed design. Each of these steps is described as follows:

- Global Routing (*axgGlobalRoute*)

Global router uses a three-dimensional array of global routing cells to model the demand and capacity of the global routing. Astro assigns nets to the global routing cells through which they pass. For each global routing cell, the routing capacity is calculated according to the blockages, pins, and routing tracks inside the cell. Astro calculates the demand for wire tracks in each global routing cell and reports the overflows, the amount of wire tracks still needed after the tool assigns nets to the available wire tracks in a global routing cell. It considers spacing and wide-wire variable routing rules, as well as shielding variable routing rules when calculating congestion.

There was no congestion while routing Matrix multiplier design. Hence no congestion maps were generated.

- Track Assignment (*axgAssignToTracks*)

Before Detail routing, Track Assignment is done to specify the tracks within each global routing cell to be used for each net. Track assignment operates on the entire design at once; it can make long routes straight and reduce the number of vias, whereas the detail router routes small area at a time.

- Detail Routing (*axgDetailRoute*)

After Track assignment, all nets are routed but not very carefully. There may be many violations particularly where the routing connects to pins. The detail router works to correct the violations and detail routing is done for the design.

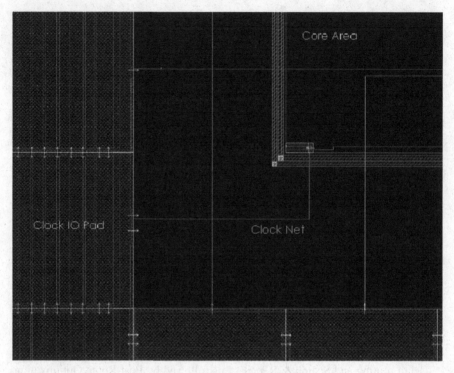

**Fig. 3.26** Snapshot of cell showing routed clock net from IO pad

- Search & Repair (*axgSearchRepair*)
  After Detail routing, Search & Repair is done on the design for searching
  DRC violations and rerouting wires in order to fix or avoid violations.

  Search & Repair is done on the Matrix multiplier design with a loop
  count of five.
- Post-Route Optimization (*astPostRouteOpt*)
  Post-Route Optimization is done to fix setup, hold, max capacitance, max
  transition violations, and maximum lengths by netlist changes and routing
  modifications at various stages of routing.

  In Matrix multiplier design, only setup slack and max capacitance violations were present. The above options were used in the optimization stage to
  eliminate the violations.

The snapshot of a section of the cell after Routing stage is shown in Fig. 3.27.

The Timing Report for the multiplier design after Routing stage is shown in
Fig. 3.28.

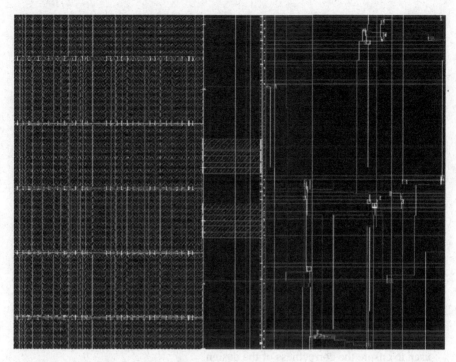

**Fig. 3.27** Snapshot of a section of the cell after routing stage

```
ASTSUM: Summary of timing analysis (w/o xtalk)
ASTSUM:      Setup (Target=0.0000)       Hold        Num      Num
ASTSUM:      Slack     Num     Total     Slack   Num MaxTrans MaxCap
ASTSUM:      0.390      0       0.0      0.072     0     0      24
@@@ Total CPU     Time  =    0:00:00
@@@ Total Elapsed Time  =    0:00:00
@@@ Peak  Memory  Used  =   96.55 MB

Timing Report OK
```

**Fig. 3.28** Timing report for matrix multiplier design after routing stage

### 3.3.2.8 Design for Manufacturability

DFM is done to address several issues to increase manufacturing yield. Before incorporating DFM for the design, DRC, ERC and LVS need to be verified to ensure error free design.

1. Design Rule Check (DRC)

   DRC is the area of Electronic Design Automation (EDA) that determines whether a particular chip design/layout satisfies a series of recommended parameters called design rules. A design rule-set specifies certain geometric and connectivity restrictions to ensure sufficient margins to account for variability in semiconductor manufacturing process, so as to ensure that most of the parts work correctly [9].

   The DRC is done for the Matrix multiplier design. For the initial round of rule check, the tool reported Notch errors. Notch filling is done for the design using the command "*geNewFillNG*". DRC is again done on the design. DRC error cell is not generated for this check, indicating that clean DRC design is obtained [9].

2. Electrical Rule Check (ERC)

   ERC involves checking a design for all well and substrate areas for proper contacts and spacing thereby ensuring correct power and ground connections [9]. ERC steps can also involve checks for unconnected inputs or shorted outputs.

3. Layout Versus Schematic (LVS)

   The LVS is the class of EDA verification software that determines whether a particular integrated circuit layout corresponds to the original schematic or circuit diagram of the design [9]. A successful DRC ensures that the layout conforms to the rules designed/required for faultless fabrication. However, it does not guarantee if it really represents the circuit desired to fabricate. Hence LVS is used to ensure the correctness of the design.

The Manufacturability issues that need to be taken care after obtaining DRC clear design are as follows:

• Gate Oxide Integrity

   The thin gate oxide may be damaged during the manufacturing process due to charge accumulation on the interconnect layers during certain fabrication steps like Plasma etching, which creates highly ionized matter to etch [9]. This is also known as *Antenna effect*. A typical Antenna effect scenario is shown in Fig. 3.29 [9].

   As length of wire increases during processing, the voltage stressing the gate oxide increases leading to Antenna effect. Antenna check rules define acceptable length of wires and also insert diodes to clamp the voltage swing. The solutions adopted to fix Antenna effect is shown in Fig. 3.30 [9].

• Via resistance and reliability

   Replacing one via contact with multiple contacts without re-routing improves both yield and timing. This advantage is due to reduction of series via resistance to parallel resistance.

• Metal erosion and Liftoff

   In Chemical Mechanical Polishing process, the wafer is made flat leaving metal tops with concave shape (Dishing). This is due to the metal being mechanically softer compared to dielectrics. Wide traces with little intervening dielectric is called *Erosion*.

**Fig. 3.29** Typical antenna effect scenario

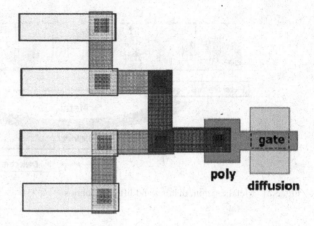

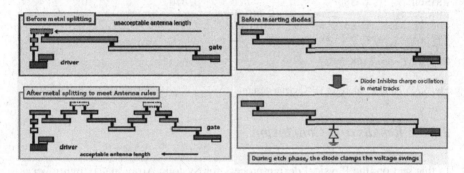

**Fig. 3.30** Solutions to fix antenna effect

Conductors and Dielectrics have different coefficients of thermal expansion. As stress builds up with temperature cycling, metal can delaminate (lift off) with time. A typical Metal erosion and liftoff case is shown in Fig. 3.31 [9].

The solution for this issue is to slot wide wires to reduce metal density. Hence minimizing stress buildup and reducing liftoff tendency.

• Metal over-etching

A narrow metal wire separated from other metal receives a higher density of etchant than closely spaced wires. Hence the narrow metal may be over-etched. This issue can be controlled by using minimum metal density rules. Filling up empty tracks with metal shapes helps in meeting minimum metal density rules. But the limitation of this solution is that no further routing or antenna fixing can be done.

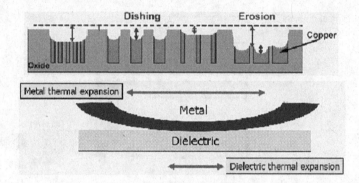

**Fig. 3.31**  Metal erosion, dishing and liftoff scenarios

```
ASTSUM: Summary of timing analysis (w/o xtalk)
ASTSUM:       Setup (Target=0.0000)              Hold          Num        Num
ASTSUM:        Slack      Num      Total      Slack    Num    MaxTrans  MaxCap
ASTSUM:        0.388       0        0.0       0.070     0        0        24
@@@ Total CPU      Time  =    0:00:00
@@@ Total Elapsed Time  =    0:00:00
@@@ Peak  Memory  Used  =   96.70 MB

Timing Report OK
```

**Fig. 3.32**  Timing report for DRC clean design

### 3.3.3   Results and Conclusion

In this section, the Physical design process for Systolic Array matrix multiplier is
carried out using Synopsys Astro design tool. The TSMC 130 nm technology is
used in the design. The design is verified for DRC errors. The Timing report for
DRC clean design obtained is shown in Fig. 3.32. The report indicates positive setup
and hold slack. The SPEF file is generated for the design to analyze and verify the
power requirements for the design.

The GDS II data file is generated for the design using "*auStreamOut*" command.
GDS II is a standard format for physical layout information. This file is used to
transport physical layout designs between different design environments. A detailed
summary report for placement and routing of the design cell is generated. The report
is analyzed for any requirement of further optimization in the design. The design is
signed off as it meets the expected requirements.

To conclude with the analysis and results obtained, there is enough scope for
enhancement and improvements in the design and verification carried out in this
section. A hierarchical design with soft and hard macros may be used to explore the
advantages of new power/ground network design flow. The technology library used
may be enhanced to 90 nm or 65 nm to experience the complexity in the design and
explore possible challenges to meet the power, area and timing requirements.

# Appendix

A. Verilog HDL code for systolic array matrix multiplier

```verilog
module matrix(clk,rst,v1,v2,v3,h1,h2,h3,
              CO11,CO12,CO13,CO21,CO22,CO23,CO31,CO32,CO33,EA,EB);

input [3:0]v1,v2,v3,h1,h2,h3;
input clk,rst;

output [7:0]CO11,CO12,CO13,CO21,CO22,CO23,CO31,CO32,CO33;
output [11:0]EA,EB;

wire [7:0]CO11,CO12,CO13,CO21,CO22,CO23,CO31,CO32,CO33;
wire [11:0]EA,EB;

wire [3:0]a12,b21;
systolic systolic_11(h1,v1,clk,rst,CO11,a12,b21);

wire [3:0]a13,b22;
systolic systolic_12(a12,v2,clk,rst,CO12,a13,b22);

wire [3:0]b23;
systolic systolic_13(a13,v3,clk,rst,CO13,EA[3:0],b23);

wire [3:0]a22,b31;
systolic systolic_21(h2,b21,clk,rst,CO21,a22,b31);

wire [3:0]a23,b32;
systolic systolic_22(a22,b22,clk,rst,CO22,a23,b32);

wire [3:0]a02,b33;
systolic systolic_23(a23,b23,clk,rst,CO23,EA[7:4],b33);

wire [3:0]a32;
systolic systolic_31(h3,b31,clk,rst,CO31,a32,EB[3:0]);

wire [3:0]a33,b02;
systolic systolic_32(a32,b32,clk,rst,CO32,a33,EB[7:4]);

systolic systolic_33(a33,b33,clk,rst,CO33,EA[11:8],EB[11:8]);

endmodule


// Sytolic Block consisting of MAC unit

module systolic(a,b,clk,rst,C,Ai,Bi);

input [3:0]a;
input [3:0]b;
input clk;
input rst;

output [7:0]C;
output [3:0]Ai;
output [3:0]Bi;

reg [7:0]C;
reg [3:0]Ai;
reg [3:0]Bi;
```

```
always@(posedge clk)
begin
    if(rst)
    begin
        Ai <= 0;
        Bi <= 0;
        C <= 0;
    end
    else
    begin
        Ai <= a;
        Bi <= b;
        C <= C + a * b;
    end
end
endmodule
```

## B. Script to generate Value Change Dump (VCD) file for matrix multiplier

```
#Commands used in ModelSim for generation of VCD file
#Create Library
Vlib Matrix

#Compile the design
Vlog Complete_Matrix.v
Vlog Systolic_Block.v
Vlog Matrix_Testbench.v

#Load the design
Vsim matrix_tb

#Add the ports to the simulation window
Add wave *

#Create and add a VCD file in the working directory
Vcd file matrix_systolic.vcd
Vcd add /matrix_t/*

#Run the test bench for the design
Run -all

#Quit the simulation
Quit -f
```

## C. Script to analyze total dynamic power using VCD and Prime Power

```
#Commands used in PrimePower for calculation of Dynamic Power

#Read Verilog netlist file
read_verilog  CB_matrix_verilog.v

#Read the design file in DB format
read_db  CB_matrix_design.db

#Read VCD file
```

```
read_vcd matrix_systolic.vcd
read_vcd -strip_path /matrix_t/* matrix_systolic.vcd

#Calculate power for the design
calculate_power

#Report the power calculated for the design
report_power
```

# References

1. Balch M (2003) Complete digital design. McGraw-Publishers, New York
2. Memory Design Examples (2007) Altera corporation. http://www.altera.com/support/examples/exm-memory.html. Accessed 15 May 2007
3. Different RAM types and its uses (2003) http://www.computermemoryupgrade.net/types-of-computer-memory-common-uses.html. Accessed 15 May 2007
4. Bhatnagar H (2002) Advanced ASIC chip synthesis using Synopsys design compiler, physical compiler and prime time, 2nd edn. Kluwer Academic Publishers, Boston
5. Lang HW, Flensurg FH (2006) Instruction Systolic Array (ISA). Institut für medieninformatik und technische informatik. http://www.iti.fh-flensburg.de/lang/papers/isa/index.htm. Accessed 15 May 2007
6. ASIC Premier (2000) LSI Logic Corporation, USA
7. Wong CK, Sarrafzadeh M (1966) An introduction to VLSI physical design. McGraw-Hill, New York
8. Astro User Guide (2005) Synopsys Inc., USA
9. Cell Based IC Physical Design and Verification with Astro (2003) National Chip Implementation Center, Taiwan

# Chapter 4
# Analog and Mixed Signal Design

The demand for Digital processing of data is seamlessly increasing for various day to day applications around us. It is because of the easier, faster and cheaper way of processing and storing data in digital format, yet efficiently. This in-turn has resulted in demand for Mixed Signal processing systems to interface with the analog and digital world. The challenges in designing a Mixed Signal system are to suppress phase noise, higher switching speeds and optimum conversion capabilities with least power dissipation. PLL, OPAMP, DAC, ADC, etc. are some of the key building blocks in an Analog and Mixed Signal System.

In this chapter a Two Stage OPAMP is designed and modeled using SPICE based on the specifications provided for 180 nm technology. The simulations are carried out using LTspice tool to extract and verify the design parameter. A layout is designed for the OPAMP. DRC and LVS debug tools are used to verify the design rules and connectivity of the layout. Parasitics are also extracted and analyzed for the design. All these processes are carried out using Cadence Virtuoso Schematic and Layout editor tool for 180 nm technology.

The prerequisite to approach this chapter would be an adequate knowledge of CMOS designs in Analog domain and basic knowledge of layout designs and SPICE modeling.

## 4.1 Schematic Design of OPAMP

### 4.1.1 Introduction

An Operational Amplifier is a DC coupled high gain electronic voltage amplifier with differential inputs and usually a single output [1]. A two stage OPAMP consists of three major blocks – Differential Amplifier stage, Gain Stage with Compensation capacitor to lower the gain at high frequencies and Buffer. An OPAMP is used in a variety of applications in linear circuit applications: Differential amplifier, inverting

V.A. Chandrasetty, *VLSI Design: A Practical Guide for FPGA and ASIC Implementations*, SpringerBriefs in Electrical and Computer Engineering, DOI 10.1007/978-1-4614-1120-8_4, © Springer Science+Business Media, LLC 2011

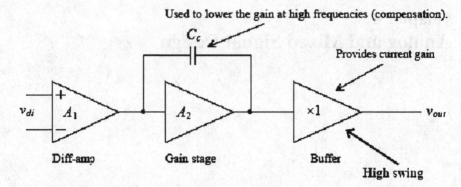

**Fig. 4.1** The functional block diagram of a two stage OPAMP

and non-inverting amplifier, Integrator, Differentiator, Comparator, Voltage follower, etc. and in non-linear circuit applications: Peak detector, logarithmic, exponential outputs, PLL, ADC, DAC, etc. The functional block diagram of a Two Stage OPAMP is shown in Fig. 4.1 [2].

### *4.1.2   Two Stage OPAMP Design*

A Two Stage OPAMP is designed and simulated in this section [2]. The design is done using SPICE modeling and the simulations are carried out using LTspice to extract and verify the design parameters against the designed values. The model file obtained from MOSIS-TSMC library for 180 nm technology [3] is used in the OPAMP modeling and simulations.

#### 4.1.2.1   Specifications

The Two Stage OPAMP is designed for TSMC 180 nm technology for the following specification:

- Open Loop Gain, $\mathbf{Av} > 100$ V/V (40 dB)
- Power Supply, $\mathbf{VDD} = -\mathbf{VSS} = 2.5$ V
- Gain Bandwidth at $-3$ dB gain, $\mathbf{f_{3db}} > 5$ MHz
- Load Capacitance, $\mathbf{C_L} = 10$ pF
- Slew Rate, $\mathbf{SR} > 10$ V/$\mu$s
- Output Voltage Swing, $\mathbf{V_{out}} = \pm 2$ V
- Input Common Mode Range, $\mathbf{ICMR} = -1$ V to $+2$ V
- Maximum Power Dissipation, $\mathbf{P_d} \leq 2$ mW
- Phase Margin, $\mathbf{\Phi_m} \geq 60°$
- Channel Length, $\mathbf{L} = 180$ nm

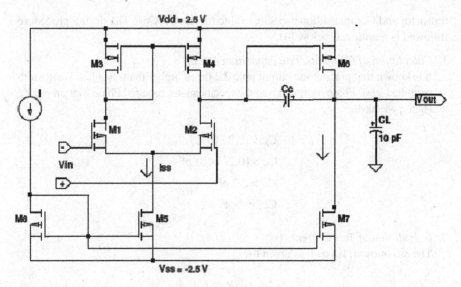

**Fig. 4.2**   The schematic diagram of two stage OPAMP

For 180 nm technology, the MOS device parameters obtained from MOSIS-TSMC fabrication process lab is as follows:
For NMOS:

- $K_n' = (\mu_n C_{ox})/2 = 177.2\ \mu A/V^2$
- $V_{tn} = 0.35\ V$
- $\lambda_n = 0.09/V$

For PMOS:

- $K_p' = (\mu_p C_{ox})/2 = -35.6\ \mu A/V^2$
- $V_{tp} = -0.39\ V$
- $\lambda_p = 0.1/V$

### 4.1.2.2   Schematic of OPAMP

The schematic diagram of Two Stage OPAMP for which aspect ratios for MOS transistors and compensation capacitance values is required to be calculated is shown in Fig. 4.2 [4].

### 4.1.2.3   Design Calculations

The Two Stage OPAMP is designed as per the specifications listed in Sect. 4.1.2.1. The end results of the design calculations are the channel width of each of the MOS

transistor and Compensation capacitor value for the OPAMP. The design procedure
followed is mentioned below [4]:

1. *Calculation of Compensation capacitance* (Cc):
   It is known that placing the output pole 2.2 times higher than the Gain Bandwidth
   permitted a 60° Phase Margin. From the specifications, required Phase Margin is 60°.
   Hence we have,

$$Cc > (2.2 / 10)C_L$$
$$Cc > (0.22) \times 10 \text{ pF}$$
$$Cc > 2.2 \text{ pF}$$
$$Cc = 3pF$$

2. *Calculation of Tail Current* (Iss):
   The tail current, Iss or $I_s$ is given by,

$$Iss = SR \times Cc$$
$$Iss = 10 \text{ V} / \mu s \times 3 \text{ pF}$$
$$Iss = 30\mu A$$

3. *Calculation of Aspect ratios* $(W/L)_3$ and $(W/L)_4$ for M3 and M4:
   The aspect ratio for M3 is calculated based on the ICMR (max) given in the
   specification.

$$(W/L)_3 = (2 \times I_5) / \left[ K_p' \left( V_{DD} - V_{in-max} - \left| V_{tp} \right| + V_{tn} \right) \right]^2$$
$$(W/L)_3 = (2 \times 30 \times 10^{-6}) / \left[ 2.5 - 2 - 0.39 + 0.35 \right]^2$$
$$(W/L)_3 = 3.98$$
$$(W/L)_3 = (W/L)_4 = 4$$

4. *Calculation of Aspect ratios* $(W/L)_1$ and $(W/L)_2$ for M1 and M2:
   The aspect ratio for M1 is calculated based on the Gain specification given.

$$Av = [2 / (\lambda n + \lambda p)] \times \left[ (2 \times K_n' \times W) / (Iss \times L) \right]^{\frac{1}{2}}$$

Given Specification, Av > 100 V/ V
Substituting and solving the values in the above equation, we get,

$$(W/L)_1 = 7.64$$
$$(W/L)_1 = (W/L)_2 = 8$$

5. *Calculation of Aspect ratios* $(W/L)_5$ and $(W/L)_8$ for M5 and M8:
The aspect ratio for M5 is calculated based on ICMR (min) specification.

$$V_{ds5} = V_{in}(min) - V_{ss} - \left[ I_5 / \left( K_n' (W/L)_1 \right) - Vtn \right]$$

Substituting the values from the specification data and previous calculations, We get,

$$V_{ds5} = 1.005 \text{ V}$$
$$(W/L)_5 = (2 \times I_5) / \left[ K_n' \times (V_{ds5})^2 \right]$$

Substituting the values in the above equation, we get,

$$(W/L)_5 = 0.34$$
$$(W/L)_5 = (W/L)_8 = 1$$

6. *Calculation of Aspect ratio* $(W/L)_6$ for M6:
The Transconductance of the input transistor M1 is given by,

$$g_{m1} = \text{(Gain Bandwidth) x (Compensation Capacitance)}$$
$$g_{m1} = 2\pi \times 5 \times 10^6 \times 3 \times 10^{-12}$$
$$g_{m1} = 94.25 \mu S$$

The Transconductance of the transistor M6 is calculated for the given specification of Phase Margin $\geq 60°$

$$gm6 \geq 10gm1$$
$$gm6 = 942.5 \mu S$$

The aspect ratio for M6 is calculated as follows:

$$(W/L)_6 = g_{m6} / \left[ K_p' \times V_{ds6}(\text{sat}) \right]$$

Substituting values in the above equation, we get,

$$(W/L)_6 = 54$$

7. *Calculation of Aspect ratio* $(W/L)_7$ for M7:
The current flowing through transistor M6 is given by

$$I_6 = (g_{m6}) / \left[ 2 \times K_p' \times (W/L)_6 \right]$$

**Table 4.1** Channel width of MOS transistors designed for 180 nm technology OPAMP

| MOS transistor | Aspect ratio (W/L) | Channel width ($\mu$m) |
| --- | --- | --- |
| M1 | 8 | 1.44 |
| M2 | 8 | 1.44 |
| M3 | 4 | 0.72 |
| M4 | 4 | 0.72 |
| M5 | 1 | 0.4 |
| M6 | 1 | 0.4 |
| M7 | 54 | 9.72 |
| M8 | 8 | 1.44 |

Substituting the values in the equation,

$$I_6 = 230\mu A$$

The aspect ratio for M7 is given by the following equation:

$$(W/L)_7 = (W/L)_5 \times (I_6/I_5)$$

Substituting values in the above equation, we get,

$$(W/L)_7 = 8$$

### 4.1.2.4  Design Calculation Results

The maximum power dissipation for the design is verified against the specification as follows:

Power Dissipation,

$$Pd(max) = (I_5 + I_6) \times (VDD + |VSS|)$$
$$Pd(max) = (30\mu + 230\mu) \times (2.5 + |-2.5|)$$
$$Pd(max) = 1.3 \text{ mW}$$

Max. power dissipation for the design is less than the specified limit of **2 mW**.

The channel width required for each of the MOS transistors for the OPAMP designed is calculated from the aspect ratios. For 180 nm process technology the channel width is tabulated as shown in Table 4.1

Other important parameters calculated in the design steps are as follows:

- Compensation Capacitance, $\mathbf{C_c = 3\ pF}$
- Load Resistance (Arbitrary value), $\mathbf{R_L = 100\ k\Omega}$
- Current flowing through M5 (Tail Current), $\mathbf{I_5 = 30\ \mu A}$
- Current flowing through M6, $\mathbf{I_6 = 230\ \mu A}$

### 4.1.2.5 Definition of Design Parameters

Definition of design parameters that are extracted from the simulation of TS-OPAMP are as follows:

1. *Open Loop Gain*: The Gain of the OPAMP for the input at positive input terminal without feedback and negative terminal input grounded
2. *Gain Bandwidth*: The frequency Bandwidth of the system at which the gain drops to −3 dB gain
3. *Phase Margin*: It is the difference measured in degrees between the absolute phase angle of OPAMP output signal and 180°
4. *Input Common Mode Range (ICMR):* The range of input voltage where the OPAMP has approximately unity gain
5. *Input Offset Voltage*: The input required to make the output of the OPAMP to zero volts
6. *Output Voltage Swing*: The range of the maximum voltage points till which the OPAMP output can swing
7. *Slew Rate*: It is the maximum rate of change of output signal at any point of time
8. *Transfer Function*: It is a function of Output of the OPAMP with respect to the Input
9. *Output Impedance*: The Impedance offered by the OPAMP at the output terminal
10. *Power Dissipation*: The total power dissipated by the OPAMP during its operation

### 4.1.2.6 Simulations and Verification

The Two Stage OPAMP designed for 180 nm process technology is simulated using LT Spice and the design specifications are verified against the extracted values [5]. The model file obtained from MOSIS-TSMC library for 180 nm technology is used in the OPAMP modeling and simulations.

- Extraction of Open Loop Gain, Gain Bandwidth and Phase Margin at 0db Gain AC analysis done to extract the above mentioned parameters. The simulation waveform obtained (Bode Plot) is shown in Fig. 4.3.

  *Configuration:* Open Loop (Extracted parameters at 0 dB gain)
  - *Gain*: 28 dB
  - *Bandwidth*: 4 MHz
  - *Phase Margin*: $(180° + \Phi) = 180° - 102° = 78°$

- Extraction of Open Loop Gain, Gain Bandwidth and Phase Margin at -3db Gain AC analysis done to extract the above mentioned parameters. The simulation waveform obtained (Bode Plot) is shown in Fig. 4.4.

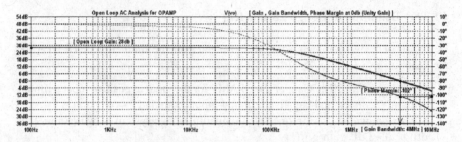

**Fig. 4.3** Simulation of TS-OPAMP to extract AC analysis parameters at 0 dB gain

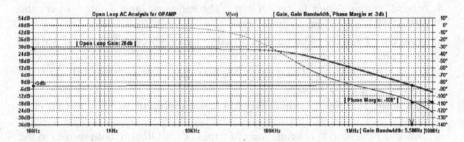

**Fig. 4.4** Simulation of TS-OPAMP to extract AC analysis parameters at 3 dB gain

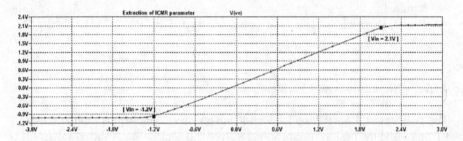

**Fig. 4.5** Simulation of TS-OPAMP to extract ICMR for the design

*Configuration*: Open Loop (Extracted parameters at −3 dB gain)
  – *Gain*: 28 dB
  – *Bandwidth*: 5.5 MHz
  – *Phase Margin*: $(180° + \Phi) = 180° − 108° = 72°$

• Extraction of ICMR
  The simulation waveform obtained to extract ICMR is shown in Fig. 4.5.

  *Configuration:* Unity Gain Feedback
  – *ICMR*: −1.2 V to +2.1 V

• Extraction of Input Offset Voltage
  The simulation waveform obtained to extract Input Offset Voltage is shown in
  Fig. 4.6.

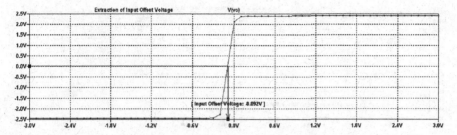

**Fig. 4.6**  Simulation of TS-OPAMP to extract input offset voltage for the design

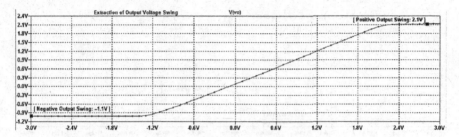

**Fig. 4.7**  Simulation of TS-OPAMP to extract output voltage Swing

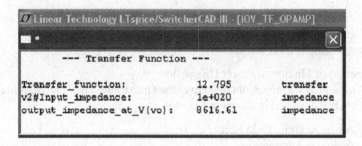

**Fig. 4.8**  Snapshot of the transfer function computed for TS-OPAMP design

*Configuration:* Open Loop
 – *IOV*: −92 mV

• Extraction of Output Voltage Swing
  The simulation waveforms obtained to extract Output Voltage Swing is shown in Fig. 4.7.

  *Configuration:* Open Loop
   – *OVS*: −1.1 V to 2.1 V

• Extraction of Transfer function and Output Impedance
  The simulation results obtained to extract Transfer function and Output Impedance of the design is shown in Fig. 4.8.

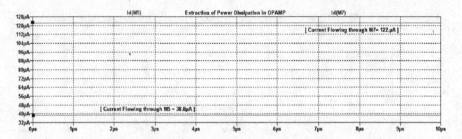

**Fig. 4.9** Simulation of TS-OPAMP to extract max. Power dissipation of the design

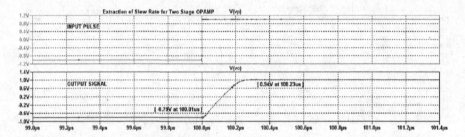

**Fig. 4.10** Simulation of TS-OPAMP to extract slew rate for the design

*Configuration:* Open Loop
- Transfer Function: 12.795
- Output Impedance: 8.6 kΩ

- Extraction of Maximum Power Dissipation
  The simulation waveform obtained to extract maximum Power Dissipation of the TS-OPAMP designed is shown in Fig. 4.9.

*Configuration*: Unity Gain Feedback
- Max. Power Dissipation,

$$P_d = (38.8\mu A + 122\mu A) \times (2.5\ V + |-2.5\ V|)$$
$$P_d = 0.804\ mW$$

- Extraction of Slew Rate
  The simulated waveform obtained to extract Slew Rate for the design is shown in Fig. 4.10.

*Configuration*: Unity Gain Feedback
- Slew Rate (SR) $= (V_2 - V_1)/(T_2 - T_1)$

$$SR = \left[0.94V - (-0.79V)\right]/(100.23\mu s - 100.01\mu s)$$
$$SR \approx 8V/\mu s$$

**Table 4.2** Comparison of design specification against results obtained for TS-OPAMP design

| Parameters | Design specification | Results obtained |
|---|---|---|
| Open loop gain (Av) | 100 V/V (40 dB) | 28 dB |
| Band width (BW) at | | |
| 0 dB | – | 4 MHz |
| 3 dB | 5 MHz | 5.5 MHz |
| Phase margin ($\Phi$) at | | |
| 0 dB | – | 72° |
| 3 dB | $\geq 60°$ | 78° |
| ICMR | –1 V to +2 V | –1.2 V to +2.1 V |
| Slew rate | 10 V/$\mu$s | 8V/$\mu$s |
| Output voltage swing | –2 V to +2 V | –1.1 V to +2.1V |
| Input offset voltage | – | –92 mV |
| Max. power dissipation | $\leq 2$ mW | 0.804 mW |
| Transfer function | – | 12.795 |
| Output impedance | – | 8.6 k$\Omega$ |

### 4.1.3   Results

The result obtained from the simulations carried out for TS-OPAMP is verified against the specification. The comparison results are tabulated as in Table 4.2.

## 4.2   Layout Design of OPAMP

### 4.2.1   Introduction

The Two Stage OPAMP designed in Sect. 4.1 is implemented to obtain the layout with optimal area and least parasitics for 180 nm technology. A schematic of TS-OPAMP is also drawn along with the layout. Cadence Virtuoso tool is used to draw schematic and layout for the design. After obtaining the layout with clean DRC and LVS, the netlist along with the parasitics is extracted with the help of the tool. Post layout simulation is carried out using this netlist to verify the design specifications.

### 4.2.2   Layout Design

In this section, the procedure for schematic and layout design of TS-OPAMP is illustrated.

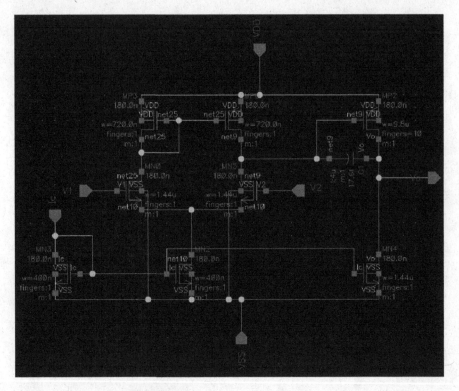

**Fig. 4.11** Schematic of TS-OPAMP

### 4.2.2.1   Schematic Design of OPAMP

The schematic design is required to carry out LVS after drawing the layout section
to verify the connectivity of the circuit. The screenshot of the schematic design of
TS-OPAMP is shown in Fig. 4.11. The components are chosen as per the designed
results available in Table 4.2. Metal plate capacitor is selected for the layout design
for compensation capacitor.

### 4.2.2.2   Layout Design of OPAMP

The Layout of OPAMP is drawn as per the schematic in Fig. 4.11. From the Table 4.2
it can be noted that MOSFET M7 has very large channel width. In order to avoid
delays and other parasitic effects caused due to large channel width, fingering is
done to break up the MOSFET into 10 MOSFETs of equal channel width [6]. The
screenshot of MOSFET M7 with finger – 10 is shown in Fig. 4.12.

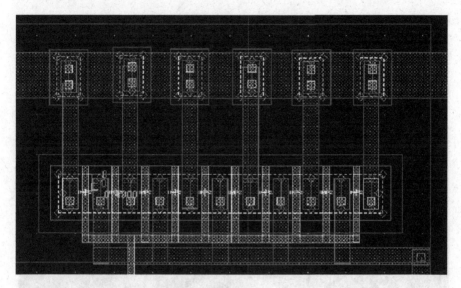

**Fig. 4.12** Screenshot of MOSFET with finger-10

Since the finger for M7 is 10, the total channel width of 9.8 μm is divided into 10 MOSFETs with channel width of 0.98 μm each. The Fig. 4.12 shows the alternate connections made to the source of MOSFET to connect it to the VDD power line. Similarly, alternate connections are done for the drain as well.

The screenshot of completed layout design of TS-OPAMP is shown in Fig. 4.13.

The completed layout of TS-OPAMP is verified for DRC. Once the layout is DRC clean, LVS is performed against the schematic to verify the connectivity of the design. LVS match is obtained for the design. The screenshot of LVS match indicator for the design is shown in Fig. 4.14.

For the LVS matched layout design, the SPICE netlist along with parasitics is extracted using RCXT tool in Cadence Virtuoso. Graphical view of the parasitics such as, resistance and capacitance in the layout design is also observed. Some of the screenshots obtained to illustrate the parasitics in the layout design are shown in the following figures.

The screenshot of the complete TS-OPAMP layout with parasitics identified is shown in Fig. 4.15.

The parasitics existing at poly of MOSFET having 10 fingers is shown in Fig. 4.16.

The parasitics identified in metal plate compensation capacitor is shown in Fig. 4.17.

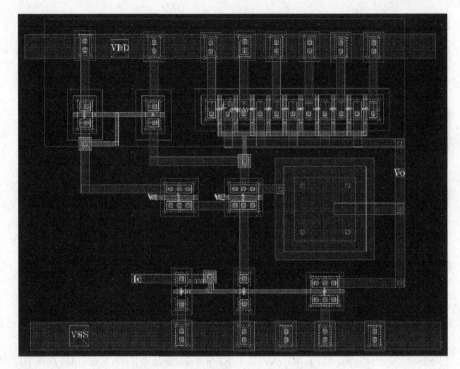

**Fig. 4.13** Screenshot of completed layout design of TS-OPAMP

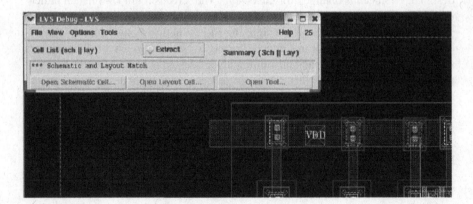

**Fig. 4.14** Screenshot of LVS match for TS-OPAMP design

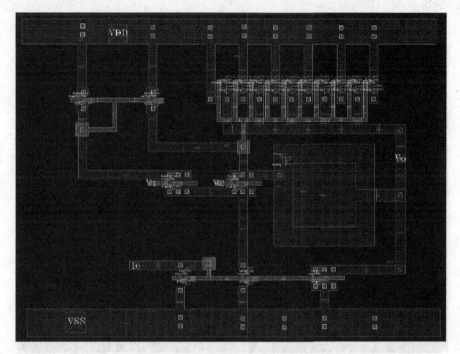

**Fig. 4.15** Screenshot of TS-OPAMP layout with parasitics identified in the design

**Fig. 4.16** Screenshot of parasitics in MOSFET layout having 10 fingers in TS-OPAMP layout design

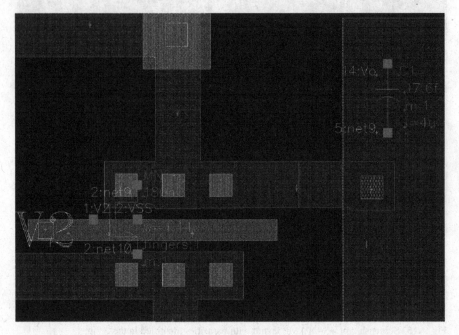

**Fig. 4.17** Screenshot of parasitics identified in layout of compensation capacitor

### 4.2.3  Summary and Results

The DRC clean and LVS match layout design of TS-OPAMP obtained have parasitics that affect the function of the design. Post layout simulation using the generated SPICE netlist for the design is carried out in LTspice to verify the specification parameters. The layout design has approximately 67 Resistances and 68 Capacitance parasitics. The area of the layout of TS-OPAMP is calculated as follows:

$$\text{Approximate Height of the Cell (H)} = 10\mu m$$
$$\text{Approximate Width of the Cell (W)} = 12\mu m$$
$$\text{Area} = H \times W = 10\mu m \times 12\mu m = 120\mu^2 m^2$$

The total area used by the TS-OPAMP layout designed cell including unused area is approximately **120 $\mu^2 m^2$**

The layout can be improved by meticulously planning the placement of MOSFETs to obtain optimized area with least parasitics. The unused area in the design can be effectively used to reduce the area metrics for the layout design. The width of the OPAMP cell is an arbitrary value as there is no reference cell with least width available. This applies also to the height of the OPAMP.

# Appendix

A. SPICE code to verify Open loop gain, Phase margin and Bandwidth using Bode plot for the OPAMP

```
* SPICE code to extract DC Gain, Phase Margin of
* Two Stage Unbuffered OPAMP @ Open Loop Configuration

.include 180nm_model.txt

M1 D13 GND SS  SS   nmos L=0.18u W=1.44u
M2 D24 Vp  SS  SS   nmos L=0.18u W=1.44u
M3 D13 D13 VDD VDD  pmos L=0.18u W=0.72u
M4 D24 D24 VDD VDD  pmos L=0.18u W=0.72u
M5 SS  CS  VSS VSS  nmos L=0.18u W=0.4u
M6 Vo  D24 VDD VDD  pmos L=0.18u W=9.72u
M7 Vo  CS  VSS VSS  nmos L=0.18u W=1.44u
M8 CS  CS  VSS VSS  nmos L=0.18u W=0.4u


C1 Vo  D24 3p
C2 Vo  GND 10p

I1 VDD CS 30u

R1 Vo  GND 100k

VDD VDD GND  2.5
VSS VSS GND -2.5

* Offset Voltage to adjust the output of OPAMP "0V" for
* Input voltage of 0V
VOF Vof  Vp 92m


* Verifying Input Offset Voltage @
*Vin Vof GND 0
*.dc Vin 0 .5 10m

* Openloop gain (Av), Phase Margin @ Unity Gain (0db) and 3db
Vin Vof GND sine(0 1 1k 0 0 0 20) AC 1
.ac lin 1k 1 10000k

.END
```

B. SPICE code to verify ICMR for the OPAMP

```
* SPICE code to extract ICMR for
* Two Stage Unbuffered OPAMP @ Unity Gain
  Feedback Configuration

.include 180nm_model.txt

M1 D13 Vo   SS   SS    nmos L=0.18u W=1.44u
M2 D24 Vp   SS   SS    nmos L=0.18u W=1.44u
M3 D13 D13  VDD  VDD   pmos L=0.18u W=0.72u
M4 D24 D24  VDD  VDD   pmos L=0.18u W=0.72u
M5 SS  CS   VSS  VSS   nmos L=0.18u W=0.4u
M6 Vo  D24  VDD  VDD   pmos L=0.18u W=9.72u
M7 Vo  CS   VSS  VSS   nmos L=0.18u W=1.44u
M8 CS  CS   VSS  VSS   nmos L=0.18u W=0.4u

C1 Vo  D24  3p
C2 Vo  GND  10p

R1 Vo  GND  10k


I1 VDD CS 30u

VDD VDD GND  2.5
VSS VSS GND -2.5

* ICMR: Range for which the Gain is unity
* Linear curve range
Vin Vp GND 0
.DC Vin -3 3 .1

.END
```

## C. SPICE code to verify Input offset voltage and output impedance for the OPAMP

```
* SPICE code to extract Input Offset Voltage, Output
  Impedence of
* Two Stage Unbuffered OPAMP @ Open Loop Configuration

.include 180nm_model.txt

M1 D13 GND SS  SS  nmos L=0.18u W=1.44u
M2 D24 Vp  SS  SS  nmos L=0.18u W=1.44u
M3 D13 D13 VDD VDD pmos L=0.18u W=0.72u
M4 D24 D24 VDD VDD pmos L=0.18u W=0.72u
M5 SS  CS  VSS VSS nmos L=0.18u W=0.4u
M6 Vo  D24 VDD VDD pmos L=0.18u W=9.72u
M7 Vo  CS  VSS VSS nmos L=0.18u W=1.44u
M8 CS  CS  VSS VSS nmos L=0.18u W=0.4u

C1 Vo  D24 3p
C2 Vo  GND 10p

I1 VDD CS 30u

R1 Vo  GND 100k

VDD VDD GND  2.5
VSS VSS GND -2.5


* Input Offset Voltage: Input voltage for which Vo is
  Zero Vin Vp GND 0
.DC Vin -3 3 0.1
.measure DC IOV when V(vo)=0

* Transfer Function and Output Impedence
*V2 Vp GND sine(0 1m 1k 0 0 0 20)
*.TF V(Vo) V2


.END
```

D. SPICE code to verify Power dissipation for the OPAMP

```
* SPICE code to extract Power Dissipated by
* Two Stage Unbuffered OPAMP @ Unity Gain
  Feedback Configuration

.include 180nm_model.txt

M1 D13 Vo   SS   SS   nmos L=0.18u W=1.44u
M2 D24 GND  SS   SS   nmos L=0.18u W=1.44u
M3 D13 D13  VDD  VDD  pmos L=0.18u W=0.72u
M4 D24 D24  VDD  VDD  pmos L=0.18u W=0.72u
M5 SS  CS   VSS  VSS  nmos L=0.18u W=0.4u
M6 Vo  D24  VDD  VDD  pmos L=0.18u W=9.72u
M7 Vo  CS   VSS  VSS  nmos L=0.18u W=1.44u
M8 CS  CS   VSS  VSS  nmos L=0.18u W=0.4u

C1 Vo   D24  3p
C2 Vo   GND  10p

R1 Vo   GND  10k

I1 VDD CS 30u

VDD VDD GND  2.5
VSS VSS GND -2.5

* Power Dissipation = (I5+I7)*(2.5 + |-2.5|)
.tran 1u 10u
.measure TRAN I5 FIND Id(M5) AT=5u
.measure TRAN I7 FIND Id(M7) AT=5u
.measure TRAN Pd   PARAM (I5+I7)*5

.END
```

E. SPICE code to verify Slew rate for the OPAMP

```
* SPICE code to extract Slew Rate
* Two Stage Unbuffered OPAMP @ Unity Gain Feedback

.include 180nm_model.txt

M1 D13 Vo   SS   SS   nmos L=0.18u W=1.44u
M2 D24 Vp   SS   SS   nmos L=0.18u W=1.44u
M3 D13 D13  VDD  VDD  pmos L=0.18u W=0.72u
M4 D24 D24  VDD  VDD  pmos L=0.18u W=0.72u
M5 SS  CS   VSS  VSS  nmos L=0.18u W=0.4u
M6 Vo  D24  VDD  VDD  pmos L=0.18u W=9.72u
M7 Vo  CS   VSS  VSS  nmos L=0.18u W=1.44u
M8 CS  CS   VSS  VSS  nmos L=0.18u W=0.4u

C1 Vo  D24  3p
C2 Vo  GND  10p

R1 Vo  GND  10k

I1 VDD CS 30u

VDD VDD GND  2.5
VSS VSS GND -2.5

* Slew Rate: [d(Vo)/Vin]/d(t) V/us
V2 Vp GND pulse(-1 1 0 .1n .1n 10u 20u)
.measure TRAN V2 FIND V(vo) AT=100.23u
.measure TRAN V1 FIND V(vo) AT=100.01u
.tran 1u 200u

.END
```

# References

1. Gayakwad RA (2000) Op-amps and linear integrated circuits, 3rd edn. Prentice-Hall, Englewood Cliffs
2. Franco S (1997) Design with operational amplifiers and analog integrated circuits, 2nd edn. McGraw-Hill Companies, Boston
3. Wafer Electrical Test Data/SPICE Model Parameters (2007) MOSIS integrated circuit fabrication service. http://www.mosis.com//Technical/Testdata/menu-testdata.html. Accessed 18 July 2007
4. Allen PE, Holberg DR (2002) CMOS analog circuit design. Oxford University Press, New York
5. Kraus AD (1991) Circuit analysis. West Publishing Company, St. Paul
6. Clein D (2000) CMOS IC layout – concepts, methodologies and tools. Newnes Publications, Boston

# About the Author

Vikram Arkalgud Chandrasetty

**Vikram Arkalgud Chandrasetty** received Bachelor Degree in Electronics and Communication Engineering from Bangalore University (India) in 2004 and Master Degree in VLSI System Design from Coventry University (UK) in 2008. He was working with Core Networks Division at Motorola India as Software Engineer (2005–2007), where he was part of the billing and call processing R&D team of Motorola Soft-Switch (MSS) for Mobile Switching Centers (MSC). He also worked for SoftJin Technologies as Senior Software Engineer (2007–2008) focusing on Electronic Design Automation (EDA) and FPGA applications design. He was involved in the design and development of Programmable Synthesis Engine (PSE) for custom FPGA architectures and structured ASICs. He was also working on software modeling and FPGA implementation of Motion Estimation algorithms for H.264 Advance Video Coder.

Mr. Vikram is currently working towards his doctoral thesis at the School of Electrical and Information Engineering, University of South Australia. He is exploring low complexity algorithms for decoding LDPC codes and investigating efficient architectures for hardware implementation. His research is mainly focused on

V.A. Chandrasetty, *VLSI Design: A Practical Guide for FPGA and ASIC Implementations,* SpringerBriefs in Electrical and Computer Engineering, DOI 10.1007/978-1-4614-1120-8, © Springer Science+Business Media, LLC 2011

implementing high performance LDPC decoders on reconfigurable devices. He has published several refereed research papers and authored two books. He is a member of Institute of Electrical and Electronics Engineers (IEEE), The Institution of Engineering and Technology (IET) and Australian Computer Society (ACS). He is also a reviewer for several international conferences and journals.